Edexcel GCSE

Mathematics A
Linear
Higher

Practice Book

Series Director: Keith Pledger
Series Editor: Graham Cumming

Authors:
Julie Bolter
Gareth Cole
Gill Dyer
Michael Flowers
Karen Hughes
Peter Jolly
Joan Knott
Jean Linsky
Graham Newman
Rob Pepper
Joe Petran
Keith Pledger
Gillian Rich
Rob Summerson
Kevin Tanner
Brian Western

Published by Pearson Education Limited, a company incorporated in England and Wales, having its registered office at Edinburgh Gate, Harlow, Essex, CM20 2JE. Registered company number: 872828

Edexcel is a registered trademark of Edexcel Limited

First published 2010

15 14
10 9 8

British Library Cataloguing in Publication Data
A catalogue record for this book is available from the British Library

ISBN 978-1-84690-084-6

Typeset by Tech-Set Ltd, Gateshead
Project management by Wearset Ltd, Boldon, Tyne and Wear
Printed in China (SWTC/08)

Disclaimer
This material has been published on behalf of Edexcel and offers high-quality support for the delivery of Edexcel qualifications.
This does not mean that the material is essential to achieve any Edexcel qualification, nor does it mean that it is the only suitable material available to support any Edexcel qualification. Edexcel material will not be used verbatim in setting any Edexcel examination or assessment. Any resource lists produced by Edexcel shall include this and other appropriate resources.

Copies of official specifications for all Edexcel qualifications may be found on the Edexcel website: www.edexcel.com

Contents

About this book

All set to make the grade!

Key points show what you need to know.

Questions match those in the Student Book.

Graded questions – so you know what you're achieving.

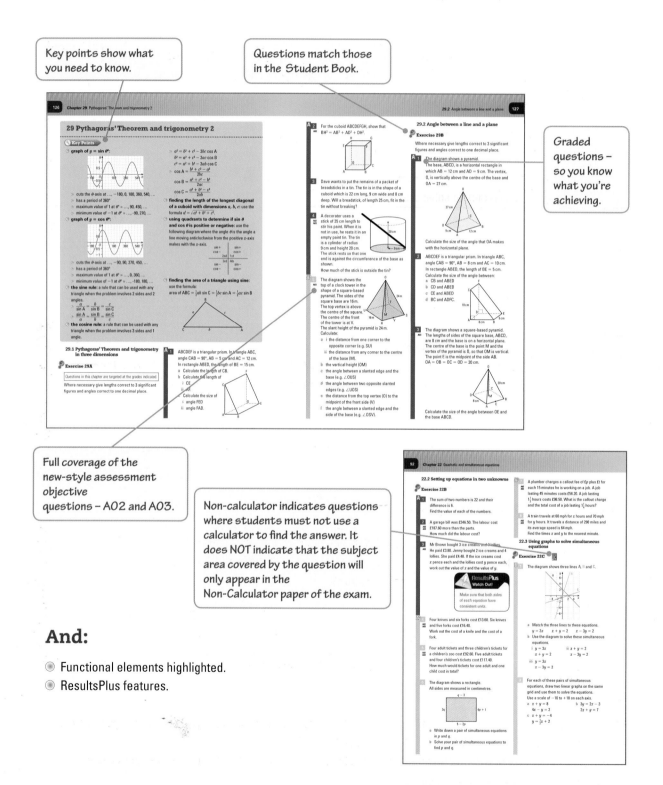

Full coverage of the new-style assessment objective questions – AO2 and AO3.

Non-calculator indicates questions where students must not use a calculator to find the answer. It does NOT indicate that the subject area covered by the question will only appear in the Non-Calculator paper of the exam.

And:

- Functional elements highlighted.
- ResultsPlus features.

Assessment Objectives

There are three types of question that are set in the exam.

Assessment Objective	What it is	What this means	Range % of marks in the exam
AO1	**Recall** and use knowledge of the prescribed content.	Standard questions testing your knowledge of each topic.	45-55
AO2	**Select** and apply mathematical methods in a range of contexts.	Deciding what method you need to use to get to the correct solution to a contextualised problem.	25-35
AO3	**Interpret** and analyse problems and generate strategies to solve them.	Solving problems by deciding how and explaining why.	15-25

The proportion of marks available in the exam varies with each Assessment Objective. Don't miss out, make sure you know how to do AO2 and AO3 questions!

What does an AO2 question look like?

D   AO2   **16**   Katie wants to buy a car.
She decides to borrow £3500 from her father. She adds interest of 3.5% to the loan and this total is the amount she must repay her father. How much will Katie pay back to her father in total?

This just needs you to (a) read and understand the question and (b) decide how to get the correct answer.

What does an AO3 question look like?

D   AO3   **17**   Rashida wishes to invest £2000 in a building society account for one year. The Internet offers two suggestions. Which of these two investments gives Rashida the greatest return?

Here you need to read and analyse the question. Then use your mathematical knowledge to solve this problem.

CHESTMAN BUILDING SOCIETY
£3.50 per month
Plus **1% bonus** at the end of the year

DUNSTAN BUILDING SOCIETY
4% per annum. Paid yearly by cheque

Quality of written communication

There will be marks in the exam for showing your working 'properly' and explaining clearly. In the exam paper, such questions will be marked with a star (*). You need to:

◉ use the correct mathematical notation and vocabulary, to show that you can communicate effectively

◉ organise the relevant information logically.

About functional elements

What does a question with functional maths look like?

Functional maths is about being able to apply maths in everyday, real-life situations.

GCSE Tier	Range % of marks in the exam
Foundation	30-40
Higher	20-30

The proportion of functional maths marks in the GCSE exam depends on which tier you are taking. Don't miss out, make sure you know how to do functional maths questions!

In the exercises...

20 The Wildlife Trust are doing a survey into the number of field mice on a farm of size 240 acres. They look at one field of size 6 acres. In this field they count 35 field mice.

a Estimate how many field mice there are on the whole farm.

b Why might this be an unreliable estimate?

> You need to read and understand the question. Follow your plan.
>
> Think what maths you need and plan the order in which you'll work.
>
> Check your calculations and make a comment if required.

ResultsPlus features

ResultsPlus features use exam performance data to highlight common pitfalls and misconceptions. ResultsPlus tips show students how to avoid errors in solutions to questions.

ResultsPlus
Watch Out!

Some students use the term average – make sure you specify mean, mode or median.

> This warns you about common mistakes and misconceptions that students frequently make.

ResultsPlus
Exam Tip

Make sure the angles add up to 360°.

> This gives exam advice, useful checks, and methods to remember key facts.

1 Number

Key Points

- **factors of a number:** whole numbers that divide exactly into the number. They always include 1 and the number itself.
- **multiples of a number:** the results of multiplying the number by positive whole numbers.
- **common multiple:** a number that is a multiple of two or more numbers.
- **prime number:** a whole number greater than 1 whose only factors are itself and 1.
- **prime factor:** a factor that is also a prime number. A number can be written as a product of its prime factors.
- **common factor:** a number that is a factor of two or more numbers.
- **lowest common multiple (LCM):** the lowest multiple that is common to two or more numbers.
- **highest common factor (HCF):** the highest factor common to two or more numbers.
- **square number:** a number that is the result of squaring a whole number.
- **square root ($\sqrt{}$):** the opposite of the square of a number. E.g. If $3 \times 3 = 9$, $\sqrt{9} = 3$.

- **cube number:** a number that is the result of cubing a whole number.
- **cube root ($^3\sqrt{}$):** the opposite of the cube of a number. E.g. If $3 \times 3 \times 3 = 27$, $\sqrt[3]{27} = 3$.
- **BIDMAS:** the order of number operations. **B**rackets, **I**ndices, **D**ivide, **M**ultiply, **A**dd, **S**ubtract.
- **inverse operations:** the inverse operation of x^y is $x^{\frac{1}{y}}$.
- **reciprocal of a number:** 1 divided by the number.
  - any number multiplied by its reciprocal $= 1$
  - zero has no reciprocal
- **index number:** a number written in the form a^n.
- **laws of indices:**
  - $a^m \times a^n = a^{m+n}$
  - $\dfrac{a^m}{a^n} = a^{m-n}$
  - $(a^m)^n = a^{m \times n}$
- **finding the square of a number:** multiply the number by itself.
- **finding the cube of a number:** multiply the number by itself and then multiply the result by the original number.

1.1 Understanding prime factors, LCM and HCF

Exercise 1A

Questions in this chapter are targeted at the grades indicated.

C **1**
A03
Can the difference between two prime numbers be a prime number?
Explain your answer.
[*Hint:* Try testing pairs of prime numbers.]

2
The number 160 can be written in the form $2^n \times 5$.
Find the value of n.

3
The number 132 can be written in the form $2^p \times q \times r$ where p, q and r are prime numbers.
Find the values of p, q and r.

C **4**
Find the HCF and LCM of the following pairs of numbers.
a 4 and 6   b 10 and 15
c 4 and 12   d 12 and 18

5
a Write 24 and 56 as products of their prime factors.
b Find the HCF of 24 and 56.
c Find the LCM of 24 and 56.

6
a Write 36 and 60 as products of their prime factors.
b Find the HCF of 36 and 60.
c Find the LCM of 36 and 60.

7
Find the HCF and LCM of the following pairs of numbers.
a 28 and 70   b 64 and 84
c 72 and 108   d 132 and 168

C 8 $c = 2^2 \times 3^3 \times 5$, $d = 2^3 \times 3^2 \times 7$

a   Find the HCF of c and d.

b   Find the LCM of c and d.

9 $s = 2^3 \times 3^3 \times 5 \times 7$, $t = 2^2 \times 5^3$

a   Find the HCF of s and t.

b   Find the LCM of s and t.

B 10   Bertrand's theorem states that 'Between any two numbers n and $2n$, there always lies at least one prime number, providing n is bigger than 1'. Show that Bertrand's theorem is true:

a   for $n = 5$ b   for $n = 12$ c   for $n = 25$.

11
A03   Frank has two flashing lamps. The first lamp flashes every 4 seconds. The second lamp flashes every 6 seconds. Both lamps start flashing together.

a   After how many seconds will they again flash together?

b   How many times in a minute will they flash together?

12
A03   A shop is going to order matching jackets and trousers. The jackets come in boxes of 5 and the trousers come in boxes of 6.

What is the smallest number of boxes of jackets and trousers that the shop must order to ensure that there is a jacket for every pair of trousers?

A 13
A03   Ben says that, if you find the average of two prime numbers, you will always get a whole number. Is Ben correct?

Give a reason for your answer.

1.2 Understanding squares and cubes

Exercise 1B

D 1   Write down:

a   the first 12 square numbers

b   the first 6 cube numbers.

ResultsPlus
Exam Tip

You need to be able to recall
- integer squares from 2×2 up to 15×15 and the corresponding square roots
- the cubes of 2, 3, 4, 5 and 10.

D 2   From each list write down all the numbers which are:

i   square numbers   ii   cube numbers.

a   50, 20, 40, 30, 4, 80, 27, 36, 25

b   64, 21, 9, 57, 60, 10, 7, 100, 48, 35, 90, 1

c   123, 25, 75, 105, 50, 125, 48, 81, 169

d   100, 175, 125, 93, 64, 75, 8, 200, 1000

Exercise 1C

1   Work out

a 3^3 b 5^3 c 4^2

d 10^2 e 14^2

2   Work out

a $\sqrt{9}$ b $\sqrt{49}$ c $\sqrt{121}$

d $\sqrt{16}$ e $\sqrt{100}$

D 3   Work out

a $(-8)^2$

b $(-3)^3$

c $(-7)^2$

d $(-2)^3$

e $(-10)^2$

ResultsPlus
Exam Tip

Remember, when multiplying or dividing:

two signs the same give a $+$

two different signs give a $-$

C 4   Work out

a $\sqrt[3]{1}$

b $\sqrt[3]{-125}$ c $\sqrt[3]{-8}$ d $\sqrt[3]{27}$ e $\sqrt[3]{-1000}$

B 5   Work out

a $3^3 + 2^2$ b $\sqrt{49} \times 5^2$

c $5^3 \times \sqrt{9}$ d $\sqrt[3]{-27} + 4^3$

e $\sqrt[3]{1000} \div \sqrt{25}$ f $10^3 \div 5^3$

g $(-1)^3 + 4^3 - (-5)^2$ h $4^3 + (-7)^2$

i $\dfrac{9^2}{3^3}$ j $2^3 \times \dfrac{\sqrt{36}}{\sqrt[3]{8}}$

k $10^3 \times \dfrac{\sqrt{225}}{\sqrt{25}}$ l $\dfrac{12^2 + \sqrt[3]{-64}}{\sqrt{16}}$

1.3 Understanding the order of operations

Exercise 1D

D 1   Work out

a $5 \times (1 + 4)$ b $3 \times 6 + 3$

c $35 \div 5 + 2$ d $24 \div (5 + 1)$

e $(7 + 3) \div -2$ f $8 + 4 \div 2$

g $24 \div (7 - 3)$ h $56 \div 8 - 2$

i $7 - (5 + 1)$ j $9 - 4 + 6$

k $5 \times 3 - 4 \times 3$ l $32 - 8 \times -5$

m $14 + 2 \times 9$ n $7 + 2 \times 4 - 12 \div 3$

o $3 - 5 \times 4 + 2$ p $(21 - 9) \times (3 + 4)$

D 2 Work out

a $(4 + 5)^2$ b $1^2 + 2^2$

c $2 \times (7 + 2)^2$ d $2 \times 4^2 + 2 \times 5^2$

e $2 \times (5 + 1)^2$ f $2 \times \sqrt{36} + 2 \times 2^3$

g $\dfrac{(17 - 3)^2}{2^2 + 3}$ h $\dfrac{3^2 - 2^2}{-5}$

B 3 Work out

a $(4 + 1)^3 \div \sqrt[3]{125}$

b $((12 - 2) \times 4) \div ((1 + 4) \times 2)$

c $9^2 + 3^3 \div \sqrt{225} + 1 \times 5$

d $(\sqrt[3]{64} + 3)^2 - \sqrt[3]{2^3 \times 5^3}$

1.4 Using a calculator

Exercise 1E

1 Work out

a $\sqrt{1024}$

b $\sqrt{42.25}$

c $\sqrt[3]{4096}$

d $\sqrt[3]{4.913}$

e $\sqrt{961}$

ResultsPlus
Watch Out!

Do not round your numbers part way through a calculation; use all the figures shown on your calculator. Only round the final answer.

D 2 Work out

a $(7.3 + 5.9) \times 1.4$ b $2.1^2 + 3.8^2$

c $(-7.8 + 5.3)^2$ d $6.5^3 + 7^2$

3 Work out, giving your answers correct to one decimal place.

a $4.7^2 \times 7.6$ b $\sqrt{35} + 5.7^3$

c $8.5^2 \div \sqrt{23}$ d $6.2^3 + \sqrt{114}$

4 Work out, giving your answers correct to three significant figures.

a $\dfrac{4.91}{3.3 - 1.82}$ b $\dfrac{8.9 \times 3.1}{1.8 \times 2.73}$

c $\dfrac{7.68 + 8.4}{4.9 - 3.74}$ d $\dfrac{5.6^2}{4.4^2 - 3.7^2}$

C 5 Work out, giving your answers correct to three significant figures.

a $\sqrt{12.49} - \dfrac{6.5}{8.1}$ b $\dfrac{6.53}{2.6} + \dfrac{1.9}{0.4}$

c $\dfrac{\sqrt{344}}{1.9 - 1.82}$ d $\left(\sqrt{\dfrac{58}{0.16}} + 576\right)^2$

C 6 Work out, giving your answers correct to three significant figures.

a $\dfrac{\sqrt{46} + 6.5^2}{18.2 + 41.3}$ b $\sqrt{\dfrac{9.8 \times 3.2}{6.9 + 8.7}}$

c $\dfrac{4.7^3}{\sqrt{8.6^2 - 1.2^2}}$ d $\dfrac{(25.3 + 8.1)^2}{\sqrt{46^2 + 27}}$

Exercise 1F

D 1 Find the reciprocal of each of the following numbers.

a 5   b 0.375   c 4.6   d $\dfrac{3}{3^3}$

1.5 Understanding the index laws

Exercise 1G

C 1 Write as a power of a single number.

a $5^4 \times 5^7$ b $2^7 \div 2^3$ c $(9^2)^4$

d $6^7 \div 6^4$ e $4^9 \times 4^3$

2 Work out

a $10^5 \times 10^3$ b $4^6 \div 4^3$ c $(3^2)^2$

d $5^5 \div 5^2$ e $2^2 \times 2$

3 Find the value of n

a $7^n \div 7^2 = 7^5$ b $5^6 \div 5^n = 5^2$

c $3^n \times 3^5 = 3^{15}$ d $9^n \times 9^7 = 9^{11}$

e $6^6 \times 6^5 = 6^n$

4 Write as a power of a single number.

a $\dfrac{8^7 \times 8^4}{8^2}$ b $\dfrac{6^6 \times 6^8}{6^4}$ c $\dfrac{4^5 \times 4^8}{4^6}$

d $\dfrac{5^{13}}{5^3 \times 5^5}$ e $\dfrac{2^5 \times 2^5}{2 \times 2^4}$

5 Work out

a $\dfrac{11^2 \times 11^5}{11^6}$ b $\dfrac{4^5 \times 4^2}{4^4}$ c $\dfrac{5^6}{5^3 \times 5}$

d $\dfrac{3^5 \times 3^5}{3^7}$ e $\dfrac{10 \times 10^9}{10^4 \times 10^3}$

B 6 Work out the value of n in the following.

a $20 = 5 \times 2^n$ b $64 = 2^n$

c $135 = 3^n \times 5$ d $24 = 3 \times 2^n$

e $162 = 2 \times 3^n$

2 Expressions and sequences

Key Points

- **variable:** something that can be changed, and is shown using a letter.
- **term:** a multiple of a variable. It can be a combination of variables and numbers such as x^3, ab, $3y^2$.
- **algebraic expression:** a collection of terms and variables.
- **like terms:** terms that use the same variable(s).
- **further laws of indices:**
  - $x^0 = 1$
  - $x^{-m} = \dfrac{1}{x^m}$
  - $x^{\frac{1}{n}} = \sqrt[n]{x}$
- **sequence:** a pattern of numbers or shapes that follow a rule.
- **terms of a sequence:** the numbers in a number sequence, or the shapes in a shape sequence.
- **term-to-term rule:** the rule of a sequence that says how to find a term from the one before it.

- **position-to-term rule:** the rule of a sequence that says how to find a term from its position in the sequence.
- **arithmetic sequence:** a sequence of numbers where the rule is to add a fixed number, known as the difference.
- **nth term:** the term in a sequence at position n from the start.
- **zero term:** the term that would be before the first term if the sequence was extended backwards.
- **simplifying algebraic expressions:** add, subtract and collect like terms, taking care with negative values. Use the laws of indices from Chapter 1.
- **substituting numbers into expressions:** replace the variables in the expression with the numeric values given.
- **finding the nth term of an arithmetic sequence:** use the result
  nth term $= n \times$ the difference $+$ the zero term.

2.1 Collecting like terms

Exercise 2A

Questions in this chapter are targeted at the grades indicated.

1   Simplify
  a $5p + 3p + 2q + q$
  b $2x + 4x + 7y - 2y$
  c $4c + d + 3d + c$
  d $3g + 4h - g - 2h$
  e $2b + a + 4a - 5b$
  f $4x - 7y - x + 3y$
  g $6j - 2k - j - 4k$
  h $6n + 4 + 2m - 3 + 2n$
  i $3e - f - 7e + 2 + 3f - 5$
  j $s + 9 - 3t + 4 - 3s - 5t$

2   Kim, James and Pusti are eating grapes. Kim has g grapes in her lunchbox. James has 7 grapes less than Kim. Pusti has 4 times as many grapes as Kim. Write down an expression, in terms of g, for the total number of these grapes.
Give your answer in its simplest form.

3   This diagram shows a rectangle.
A03

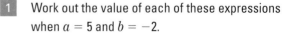

Write down an expression, in terms of x and y, for the perimeter of this rectangle.
Give your answer in its simplest form.

2.2 Using substitution

Exercise 2B

1   Work out the value of each of these expressions when $a = 5$ and $b = -2$.
  a $2a + b$ b $a - b$
  c $3a + 4b + 6$ d $5a - 3 - 2b$

2   Work out the value of each of these expressions when $x = -3$, $y = 2$ and $z = -4$.
  a $x + y + z$ b $3x + 2y + 4z$
  c $4x - y + 2z$ d $x - 6 - 3y - z$
  e $5x - 2y^2$ f $2x^2 + y^2 + 2z^2$

2.3 Using the index laws

Exercise 2C

ResultsPlus
Watch Out!

Group like terms together before attempting to use the laws of indices.

D **1** Simplify
   a $a \times a \times a \times a$ b $2f \times f \times 3f$
   c $4h \times 3h \times 5h$

C **2** Simplify
   a $k^4 \times k^5$ b $t^4 \times t$ c $c - c^6$
   d $x^4 \times x^5 \times x^3$

3 Simplify
   a $3d^3 \times 5e^4$ b $4y \times 4y^3$ c $u^9 \times 6u^3$
   d $5z^3 \times 4z$

4 Simplify
   a $5x^2y^3 \times 3x^4y^4$ b $4uv^3 \times 3u^4v^5$
   c $c^2d^3 \times 9c^3d$ d $4mn^6 \times 5mn^4$
   e $4r^2s \times 5rs^2 \times 2rs^2$

Exercise 2D

C **1** Simplify
   a $x^{10} \div x^5$ b $t^7 \div t$ c $\dfrac{h^5}{h^4}$ d $k^4 \div k^2$

2 Simplify
   a $8u^5 \div 2u^2$ b $24z^7 \div 6z^3$
   c $14w^6 \div 7w^4$ d $\dfrac{36g^5}{9g}$

B **3** Simplify
   a $18y^5z^6 \div 6y^2z^3$ b $30a^4b^3 \div 5a^2b$
   c $\dfrac{12u^4v^9}{6u^3v^4}$ d $\dfrac{3q^4 \times 8q^3}{4q^3}$
   e $\dfrac{8bc^5 \times 5b^2c}{4bc^5}$

Exercise 2E

C **1** Simplify
   a $(k^9)^3$ b $(s^4)^6$ c $(x^3)^2$ d $(a^5)^6$

2 Simplify
   a $(3w^3)^4$ b $(2x^2)^5$ c $(10y^4)^3$ d $\left(\dfrac{z^6}{4}\right)^3$

B **3** Simplify
   a $(2a^7b)^3$ b $(4g^4h^5)^3$ c $(9k^5l^8)^2$
   d $\left(\dfrac{7m^3n^6}{11m^5n}\right)^2$

2.4 Fractional and negative powers

Exercise 2F

ResultsPlus
Exam Tip

Remember that a negative power just means 'one over' or 'the reciprocal of'.

B **1** Simplify
   a w^{-2}
   b $(x^3)^{-1}$
   c y^{-1}
   d $(z^2)^{-1}$

2 Simplify
   a $(g^4)^{-2}$ b $(h^3)^{-4}$ c $(j^{-1})^{-3}$ d $(k^{-2})^{-1}$

A **3** Simplify
   a $(a^3b^7)^0$ b $(2c^5d^5)^0$ c $(5e^4f^4)^{-1}$
   d $(3x^2y)^{-3}$ e $\left(\dfrac{2x^2y}{3z^3}\right)^{-2}$

Exercise 2G

A **1** Simplify
   a $(36x^6)^{\frac{1}{2}}$ b $(16d^4)^{\frac{1}{4}}$ c $(64e^3f^{-2})^{\frac{1}{3}}$
   d $(64x^2y^3)^{\frac{1}{2}}$

2 Simplify
   a $(c^8)^{-\frac{1}{2}}$ b $(8g^6)^{-\frac{1}{3}}$ c $(32p^2q^5)^{-\frac{1}{5}}$
   d $(x^4y^6)^{-\frac{1}{4}}$

2.5 Term-to-term and position-to-term definitions

Exercise 2H

Find a the term-to-term rule,
      b the next two terms, and
      c the 10th term for each of the following
         number sequences.

1	1	5	9	13
2	-2	4	10	16
3	26	17	8	-1
4	6	9	15	24
5	0	2	6	14

2.6 The nth term of an arithmetic sequence

Exercise 2I

C **1** Write down
   i   the difference between consecutive terms
   ii the zero term for each of the following
       arithmetic sequences.
       a 1, 3, 5, 7, 9, ...   b -2, 1, 4, 7, 10, ...
       c 19, 13, 7, 1, -5, ...

C **2** Here are the first five terms of an arithmetic sequence: 1, 6, 11, 16, 21, ...

    a Write down, in terms of n, an expression for the nth term of this arithmetic sequence.

    b Use your answer to part a to work out the   i 12th term,   ii 50th term.

3 Here are the first four terms of an arithmetic sequence: 4, 9, 14, 19, ...

    a Write down, in terms of n, an expression for the nth term of this arithmetic sequence.

    b Use your answer to part a to work out the   i 15th term,   ii 100th term.

4 Here are the first five terms of an arithmetic sequence: 43, 36, 29, 22, 15, ...

    a Write down, in terms of n, an expression for the nth term of this arithmetic sequence.

    b Use your answer to part a to work out the   i 20th term,   ii 200th term.

C **5** A03 The first four terms of an arithmetic sequence are 17, 22, 27, 32, ...

Explain why the number 105 cannot be a term of this sequence.

6 * A03 The first five terms of an arithmetic sequence are 8, 14, 20, 26, 32, ...

Jacob says that 343 is a term of this sequence. Jacob is wrong. Explain why.

3 Fractions

Key Points

- **improper fraction:** a fraction where the numerator is greater than the denominator.
- **equivalent fractions:** fractions that represent the same quantity but use different denominators. E.g. $\frac{1}{2} = \frac{2}{4}$.
- **mixed number:** a number with a whole number part and a proper fraction part.
- **adding and subtracting fractions:** use equivalent fractions to get fractions with a common denominator. Add or subtract the numerators.

- **adding and subtracting mixed numbers:** add or subtract the whole numbers, then add or subtract the fractions separately.
- **multiplying fractions:** convert any mixed numbers to improper fractions, simplify if possible, and then multiply the numerators and multiply the denominators.
- **dividing fractions:** convert any mixed numbers to improper fractions, invert the dividing fraction and then multiply.

3.1 Adding and subtracting fractions and mixed numbers

Exercise 3A

Questions in this chapter are targeted at the grades indicated.

Give each answer as a fraction in its simplest form.

D **1** Work out

a $\frac{6}{12} + \frac{5}{12}$ b $\frac{2}{9} + \frac{5}{9}$ c $\frac{8}{15} + \frac{3}{15}$

d $\frac{5}{10} + \frac{2}{10}$

2 Work out

a $\frac{2}{5} + \frac{1}{2}$ b $\frac{1}{3} + \frac{3}{7}$ c $\frac{5}{7} + \frac{1}{5}$

d $\frac{3}{9} + \frac{1}{3}$ e $\frac{1}{5} + \frac{3}{4}$ f $\frac{1}{2} + \frac{4}{9}$

g $\frac{1}{9} + \frac{1}{6}$ h $\frac{3}{20} + \frac{2}{5}$

3 Work out

a $\frac{1}{2} - \frac{1}{8}$ b $\frac{1}{3} - \frac{1}{6}$ c $\frac{7}{8} - \frac{3}{5}$

d $\frac{8}{9} - \frac{1}{3}$ e $\frac{3}{4} - \frac{3}{8}$ f $\frac{2}{3} - \frac{1}{6}$

g $\frac{17}{20} - \frac{1}{4}$ h $\frac{5}{6} - \frac{2}{9}$

4 Work out **a–h**, giving each answer as a fraction or a mixed number in its simplest form.

a $\frac{3}{5} + \frac{7}{8}$ b $\frac{3}{4} + \frac{2}{5}$ c $\frac{5}{9} - \frac{1}{3}$

d $\frac{7}{10} + \frac{3}{4}$ e $\frac{5}{6} + \frac{7}{10}$ f $\frac{3}{4} + \frac{7}{12}$

g $\frac{3}{5} - \frac{1}{2} + \frac{5}{20}$ h $\frac{5}{9} + \frac{2}{3} - \frac{4}{18}$

Exercise 3B

C **1** Work out

a $3\frac{1}{4} + 2\frac{1}{2}$ b $3\frac{3}{5} + 1\frac{1}{2}$

c $5\frac{3}{8} + 6\frac{2}{5}$ d $9\frac{4}{7} + 4\frac{5}{9}$

2 Colin and Jane hiked $2\frac{1}{4}$ km. They stopped for a picnic and then hiked a further $3\frac{2}{3}$ km to a campsite.
What is the total distance that they hiked?

3 Some electrical equipment weighs $1\frac{2}{5}$ kg. The packaging it comes in weighs $\frac{5}{7}$ kg.
What is the total weight of the equipment and its packaging?

Exercise 3C

C **1** Work out

a $3\frac{1}{2} - 1\frac{3}{4}$ b $4\frac{5}{6} - 2\frac{1}{2}$

c $7 - 3\frac{3}{4}$ d $6 - 2\frac{3}{5}$

2 Work out

a $3\frac{3}{4} - 1\frac{1}{2}$ b $4\frac{1}{4} - 2\frac{1}{3}$

c $5\frac{3}{8} - 1\frac{4}{5}$ d $6\frac{7}{12} - 3\frac{1}{3}$

3 A sack containing balls has a total weight of $6\frac{2}{3}$ kg. The empty sack has a weight of $1\frac{1}{6}$ kg.
What is the weight of the balls?

4 A hedge is $5\frac{1}{2}$ feet tall. Calvin trims off the top $2\frac{3}{10}$ feet. How tall is the hedge now?

3.2 Multiplying fractions and mixed numbers

Exercise 3D

1 Work out

a $\frac{1}{5} \times \frac{1}{3}$ b $\frac{3}{4} \times \frac{3}{7}$ c $\frac{15}{17} \times \frac{1}{5}$

d $\frac{5}{12} \times \frac{4}{15}$ e $\frac{2}{5} \times \frac{2}{7}$ f $\frac{3}{8} \times \frac{3}{5}$

g $\frac{6}{28} \times \frac{7}{15}$ h $\frac{5}{36} \times \frac{27}{45}$

2 Work out

a $2 \times \frac{1}{5}$ b $3 \times \frac{1}{2}$ c $\frac{11}{20} \times 6$

d $\frac{3}{5} \times 15$

3 Work out

a $\frac{3}{5}$ of 45 litres   b $\frac{4}{9}$ of 18 kg

c $\frac{3}{8}$ of 32 m   d $\frac{7}{10}$ of 15 km

4 Narendra planted 48 tulip bulbs. When the tulips grow, $\frac{5}{6}$ of them are red.
How many tulips are red?

5 A sunflower is 150 cm tall. A week later, it has grown $\frac{1}{4}$ taller. How many cm has it grown?

6 Work out

a $2\frac{3}{4} \times \frac{1}{3}$ b $1\frac{3}{5} \times \frac{1}{4}$ c $3\frac{3}{4} \times 2\frac{7}{10}$

d $1\frac{2}{3} \times 3\frac{1}{5}$ e $1\frac{2}{3} \times 1\frac{1}{4}$ f $3\frac{1}{2} \times 1\frac{3}{4}$

g $5\frac{4}{7} \times 2\frac{3}{5}$ h $8\frac{1}{3} \times 3\frac{3}{10}$

7 It takes Mary $1\frac{1}{4}$ minutes to swim a length of a swimming pool.
How long will it take her to swim $5\frac{1}{2}$ lengths?

8 A bag of sand weighs $2\frac{1}{4}$ lbs.
Work out the weight of $12\frac{1}{2}$ bags.

3.3 Dividing fractions and mixed numbers

Exercise 3E

1 Work out

a $\frac{5}{6} \div 3$ b $\frac{5}{8} \div 3$ c $\frac{4}{5} \div \frac{7}{10}$

d $\frac{9}{16} \div \frac{1}{8}$ e $\frac{3}{4} \div \frac{1}{2}$ f $\frac{3}{5} \div \frac{1}{4}$

g $\frac{20}{21} \div \frac{8}{27}$ h $\frac{25}{32} \div \frac{5}{8}$

2 Work out

a $3\frac{1}{3} \div 7$ b $2\frac{3}{5} \div \frac{9}{10}$ c $3\frac{3}{4} \div 2\frac{2}{5}$

d $4\frac{2}{3} \div 3\frac{8}{9}$ e $4\frac{1}{2} \div \frac{3}{4}$ f $3\frac{1}{9} \div \frac{2}{3}$

g $7\frac{1}{2} \div 2\frac{1}{4}$ h $3\frac{1}{12} \div 1\frac{3}{8}$

3 A sack of nuts weighs $12\frac{2}{3}$ kg. The nuts are repackaged into small bags, each weighing $\frac{2}{3}$ kg. How many small bags can be filled from the sack?

4 A large hall is $12\frac{3}{5}$ m wide. How many square floor tiles of length $\frac{7}{10}$ m will fit the room's width?

5 George is using tins of paint that can cover $2\frac{3}{5}$ m² each. How many tins will George need to paint a $24\frac{1}{5}$ m² area?

3.4 Fraction problems

Exercise 3F

1 Sarah spends $\frac{1}{2}$ of her birthday money on some shoes and $\frac{1}{4}$ of her birthday money on a dress.
a What fraction of her money does Sarah spend on the shoes and dress altogether?
b What fraction of her money is left?

2 $\frac{5}{7}$ of bag of mixed frozen vegetables are peas. The total weight of the vegetables is 630 g.
What is the weight of the peas in the bag?

3 Tickets for a charity event cost £12 each. $\frac{3}{5}$ of the £12 goes to the charity and the rest covers the cost of hosting the event.
How much of the £12 goes to the charity?

4 A satnav usually costs £156. A sale reduces all prices by $\frac{3}{8}$.
Work out the sale price of the satnav.

5 A college has 2350 students.
1050 of the students are male.
$\frac{3}{5}$ of the male students are aged between 16 and 18. $\frac{3}{4}$ of the female students are aged between 16 and 18.
How many students are not aged between 16 and 18?

6 Josh says that only $\frac{2}{5}$ of his fellow workers are women. There are 42 workers in his department.
Explain why Josh cannot be right.

7 Mei Yin cycles $1\frac{3}{4}$ km to the library and then cycles $2\frac{1}{6}$ km to her friend's house.
How far has Mei Yin cycled altogether?

C 8 $\frac{2}{3}$ of a circle is shaded. $\frac{3}{5}$ of the shaded part is shaded green.
What fraction of the whole circle is shaded green?

9 A baker has combined equal quantities of flour, sugar and butter. The total weight of the mixture is $1\frac{7}{8}$ kg.
Work out the weight of each ingredient.

C 10 In a magazine, $\frac{3}{5}$ of the pages have adverts on. Given that 48 pages have adverts on, work out the number of pages in the magazine.

11 Adam spent $\frac{1}{3}$ of his pocket money on downloading music. He put $\frac{2}{5}$ of his pocket money into a savings account and spent the rest on a ticket for a concert. His ticket cost £12.80. How much pocket money did he get?

4 Decimals and estimation

- **decimal point:** separates the whole number part from the part that is less than 1.
- **terminating decimal:** a decimal which ends. E.g. 0.34, 0.276 and 5.089.
- **recurring decimal:** a decimal in which one or more digits repeat. E.g. 0.11111…, 0.563563563…, and 8.5644444… .
- **common fraction-to-decimal conversions:**

Decimal	0.01	0.1	0.2	0.25	0.5	0.75
Fraction	$\frac{1}{100}$	$\frac{1}{10}$	$\frac{1}{5}$	$\frac{1}{4}$	$\frac{1}{2}$	$\frac{3}{4}$

- **upper bound of a number:** the highest value which rounds down to that number.
- **lower bound of a number:** the lowest value which rounds up to that number.
- **deducing if a fraction is a terminating or recurring decimal:** write the fraction in its simplest form and write the denominator of the fraction in terms of its prime factors. If these prime factors are only 2s and 5s then the fraction will convert to a terminating decimal. If not it will be recurring.
- **converting a fraction to a decimal:** divide the numerator by the denominator.
- **adding and subtracting decimals:** keep the decimal points in line so that the place values match.

- **multiplying decimals:** do the multiplication with whole numbers and then decide on the position of the decimal point. The total number of decimal places in the answer is the same as that in the question.
- **dividing decimals:** multiply both numbers by 10, 100, 1000 etc. until you are dividing by a whole number.
- **rounding to a given number of decimal places (d.p.):** count this number of decimal places from the decimal point. If the next digit is greater than or equal to 5, round up. If not, leave off this digit and any that follow.
- **rounding to a given number of significant figures (s.f.):** count this number of digits from the first non-zero digit. If the next digit is 5 or more then round up.
- **estimating an answer:** round each number to 1 significant figure and then calculate.
- **converting a recurring decimal to a fraction:** use the following method:
  - let x equal the recurring decimal
  - multiply both sides of this equation by 10 if 1 digit recurs, 100 if 2 digits recur, and so on
  - subtract the original equation by the new equation
  - rearrange to find x as a fraction

4.1 Conversion between fractions and decimals

⚙ **Exercise 4A**   🖩

Questions in this chapter are targeted at the grades indicated.

1   Write each of the set of numbers in order, starting with the smallest.

$\frac{7}{10}$, 0.6, 0.65, $\frac{66}{100}$, 0.76

D 2   By writing the denominator in terms of its prime factors, decide whether these fractions will convert to recurring or terminating decimals.

a $\frac{7}{40}$ b $\frac{15}{32}$ c $\frac{6}{45}$ d $\frac{11}{42}$ e $\frac{8}{125}$ f $\frac{31}{60}$

D 3   Marlon says that $\frac{19}{52}$ can be converted to a terminating decimal.

Iman says that the fraction converts to a recurring decimal. Who is correct?

You must give a reason for your answer.

4.2 Carrying out arithmetic using decimals

⚙ **Exercise 4B**   🖩

1   Work out
   a 0.3 × 0.5   b 0.009 × 0.3
   c 0.7 × 0.06   d 0.09 × 0.08

2   Work out

  a 5.24×0.5 b 3.37×0.4
  c 0.291×0.06 d 3.15×0.9
  e 3.1×2.4 f 0.63×4.1
  g 0.064×0.37 h 0.082×6.1

3   Work out the cost of 0.8 kg of parsnips at 65p per kilogram.

4   Work out the cost of 1.4 m of ribbon at £1.39 per metre.

5   Work out

  a $14 \div 0.2$ b $2.4 \div 0.3$
  c $19.5 \div 0.03$ d $18 \div 0.4$
  e $7 \div 0.2$ f $9.12 \div 0.003$
  g $0.042 \div 0.7$ h $0.008\,37 \div 0.09$

6   Work out

  a $52.05 \div 0.15$ b $150.5 \div 0.25$
  c $0.8904 \div 0.042$ d $7.533 \div 0.31$

7   Six people share £156.78 equally.
    Work out how much each person will get.

8   Charlotte makes 4.2 litres of orange squash.
    She pours it into 0.6 litre bottles.
    How many bottles can Charlotte fill?

4.3 Rounding and decimal places

Exercise 4C

1   Write the following numbers correct to 1 decimal place. (1 d.p.)

  a 3.68   b 6.56   c 61.938
  d 0.076   e 0.96

2   Write the following numbers correct to 2 decimal places. (2 d.p.)

  a 4.766   b 8.0285   c 0.135
  d 4.054   e 0.067

3   Write the following numbers correct to 3 decimal places. (3 d.p.)

  a 7.3855   b 0.0894   c 6.0392
  d 4.0057   e 0.0298

4.4 Significant figures

Exercise 4D

1   Write the following numbers correct to 2 significant figures.

  a 4869   b 332.6   c 25.58
  d 7.49   e 6.062   f -0.6327

2   Write the following numbers correct to 3 significant figures.

  a 3495   b 18.96   c 3.195
  d 2.0889   e 0.020 95

3   Write the following numbers correct to 1 significant figure.

  a 4398   b 33.75   c 4.006
  d 6.73   e 29.7   f 0.954

4.5 Estimating calculations by rounding

Exercise 4E

D 1   Estimate the value of the following calculations.

  a 79×68 b 122×78 c 395×29
  d 5897×28 e 588×209

2   Work out estimates for each of the following calculations.

  a $779 \div 1.8$ b $9.7 \div 4.8$ c $67.2 \div 7.3$
  d $221.5 \div 41$ e $588 \div 96.4$

3   Work out estimates for each of the following.
A03   In each case state whether your answer is an overestimate or an underestimate of the true answer.

  a 179×37 b $18.7 \div 4.3$
  c $74.8 \div 6.83$ d $37.8 \times 1.79 \times 9.9$

C 4   Work out estimates for each of the following calculations.
    State whether your answer is an underestimate or an overestimate.

  a $\dfrac{55.3 \times 19.3}{21.2}$ b $\dfrac{4621}{11.7 \times 5.4}$

  c $\dfrac{849}{1.8 \times 3.5}$ d $\dfrac{23.7 \times 31.2}{18.6 \times 9.501}$

5   Work out an estimate for the value of 1.8×32.9^2.

4.6 Estimating calculations involving decimals

Exercise 4F

1 Work out estimates for the values of
a 5.8×0.27
b 0.66×0.21
c 24.5×0.217
d 0.095×0.013

2 Work out estimates for the values of
a $\dfrac{21.49}{3.7}$
b $\dfrac{58.2}{5.43}$
c $\dfrac{113.4}{1.75}$
d $\dfrac{0.83}{0.21}$
e $\dfrac{1.709}{4.09}$

3 Work out estimates for the values of the following.
State whether your answer is an overestimate or an underestimate.
a $\dfrac{3.8 \times 7.5}{0.21}$
b $\dfrac{5.3 \times 11.9}{0.48}$
c $\dfrac{22.6 \times 3.3}{0.26}$
d $\dfrac{182 \times 19.3}{0.84}$

4 $h = \dfrac{V}{\pi r^2}$, $V = 0.027$, $r = 0.11$. Work out an estimate for the value of h. State whether your answer is an overestimate or an underestimate.

5 $V = LWH$, $L = 0.077$, $W = 0.034$, $H = 981$. Work out an estimate for the value of V.

4.7 Manipulating decimals

Exercise 4G

1 Given that $6.3 \times 2.7 = 17.01$ work out
a 63×27
b 630×2.7
c 0.63×27
d 0.63×0.027

2 Given that $67.87 \div 2.75 = 24.68$ work out
a $6787 \div 2.75$
b $6.787 \div 2.75$
c $0.6787 \div 2.75$
d $0.6787 \div 27.5$

3 Given that $\dfrac{34.1 \times 4.2}{6.2} = 23.1$ work out
a $\dfrac{34.1 \times 42}{6.2}$
b $\dfrac{341 \times 42}{6.2}$
c $\dfrac{34.1 \times 42}{62}$
d $\dfrac{341 \times 42}{62}$

4 Given that $24 \times 55 = 1320$ work out
a 0.24×550
b $1320 \div 5.5$
c $13.20 \div 0.24$
d $1320 \div (24 \times 27.5)$

5 Given that $882 \div 32 = 27.6$ work out
a $8.82 \div 320$
b $882 \div 2.76$
c $8.82 \div 276$
d $882 \div (3.2 \times 2.76)$

6 Given that $\dfrac{1874}{1.4^2} = 956$ work out
a $\dfrac{1874}{14^2}$
b $\dfrac{18.74}{1.4^2}$
c $\dfrac{187.4}{0.14^2}$
d $\dfrac{937}{140^2}$

4.8 Converting recurring decimals to fractions

Exercise 4H

Convert each recurring decimal to a fraction.
Give each fraction in its simplest form.
Do **not** use a calculator. You must use algebra.

1 $0.111\,11\ldots$
2 $0.363\,636\ldots$
3 $0.162\,162\ldots$
4 $0.2\dot{7}$
5 $0.2\dot{1}\dot{6}$
6 $0.0\dot{3}$
7 $0\dot{1}3\dot{6}$
8 $0.1\dot{8}\dot{2}$
9 $0.1\dot{6}$
10 $6.8\dot{2}$
11 $3.08\dot{6}$
12 $7.86\dot{1}$

4.9 Upper and lower bounds of accuracy

Exercise 4I

ResultsPlus
Exam Tip

Remember that the upper bound is the same distance above x as the lower bound is below x.

1 Write down:
i the upper bound and
ii the lower bound of these numbers.
a 73 (2 significant figures)
b 73.0 (3 significant figures)
c 73.00 (4 significant figures)

2 Write down:
i the upper bound and
ii the lower bound of these numbers.
a 0.8 (1 decimal place)
b 0.80 (2 decimal places)
c 0.08 (2 decimal places)

A 3 The distance between two towns is 129 km correct to the nearest km. Write down:

a the upper bound

b the lower bound of the distance.

Give your answers in km.

4 The length of a centipede is 7.6 cm correct to the nearest millimetre. Write down:

a the upper bound

b the lower bound of the length of the centipede.

Give your answers in mm.

5 The amount of water in a water butt is 52.0 litres correct to the nearest tenth of a litre.
Write down:

a the upper bound

b the lower bound of the amount of water in the water butt.

Give your answers in litres.

6 The length of a table is 2 metres correct to the nearest cm. Write down:

a the upper bound

b the lower bound of the length of the table.

Give your answers in metres.

A* 4 $y = 0.5at^2$, $a = 4.5$ (1 d.p.), $t = 6.8$ (1 d.p.).
A02

a Find the lower bound of y.

b Find the upper bound of y.

5 $p = \dfrac{l^2}{w}$, $l = 6.44$ (2 d.p.) and $w = 5.45$ (2 d.p.).
A02

a Calculate the lower bound of p.

b Calculate the upper bound of p.

6 Jim was at the racing track. He estimated that
A02 the length of the track he could see was 100 m
A03 correct to the nearest 10 metres. He timed a car as taking 5 seconds to the nearest second along the track he could see.
Work out the lower bound and the upper bound for the average speed of the car.
Give your answers correct to 1 decimal place.

7 $y = a\cos x°$, $a = 4.2$ (1 d.p.) and $x = 60°$ (2 s.f.).
A03 Show that the lower bound of y is 2.04 (3 s.f.) and find the upper bound of y.

8 a Expand $(a + b)^2$.
A03
$y = n^2$
n is written correct to the nearest whole number.

b Show that the upper bound of y is $n^2 + n + \dfrac{1}{4}$.

c Find a similar expression for the lower bound of y.

4.10 Calculating the bounds of an expression

Exercise 4J

A 1 $x = 4.0$ (1 d.p.), $y = 5.2$ (1 d.p.).
A02 Work out the lower bounds of

a $x + y$ b xy

2 $m = 3.5$ (1 d.p.), $n = 5.7$ (1 d.p.).
A02 Work out the upper bounds of

a $m + n$ b mn

3 $a = 3.44$ (2 d.p.), $b = 1.85$ (2 d.p.).
A02 Work out the upper bounds of

a $a + b$ b ab c a^2

5 Angles and polygons

Key Points

- **angles on parallel lines:**
  - **corresponding angles are equal**
  - **alternate angles are equal**

- **exterior and interior angles:**
  - angle *e* is an exterior angle
  - angle *i* is an interior angle

- **angle properties of triangles and quadrilaterals:**
  - the exterior angle of a triangle equals the sum of the interior angles at the other two vertices
  - the angel sum of a quadrilateral is 360°
  - opposite angles of a parallelogram are equal

- **angles of elevation and depression:**
  - angle of elevation from point A to point B is the angle of turn above the horizontal to look directly from B to A
  - angle of depression of point B from point A is the angle of turn below the horizontal to look directly from A to B

- **bearing:** a way to describe direction as an angle measured clockwise from North and written as a 3 figure number.
- **polygon:** a 2D shape with straight sides.
- **regular polygon:** a polygon with equal sides and equal interior angles.

Name	Number of Sides	Shape
Triangle	3	
Quadrilateral	4	
Pentagon	5	
Hexagon	6	
Heptagon	7	
Octagon	8	
Decagon	10	

- **angles of a polygon:**
  - **the sum of the interior angles of a polygon $= (n - 2) \times 180°$ where n is the number of sides**
  - **interior angle of a polygon + exterior angle of a polygon $= 180°$**
  - **the sum of the exterior angles of a polygon $= 360°$**

5.1 Angle properties of parallel lines

Exercise 5A

Questions in this chapter are targeted at the grades indicated.

In questions 1−6 find the size of each lettered angle. Give reasons for your answers.

D **1**

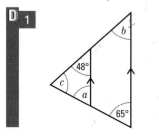

2

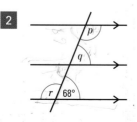

D **3**   **4**

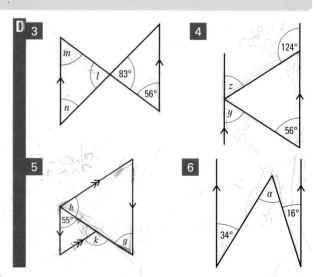

5   **6**

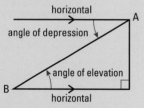

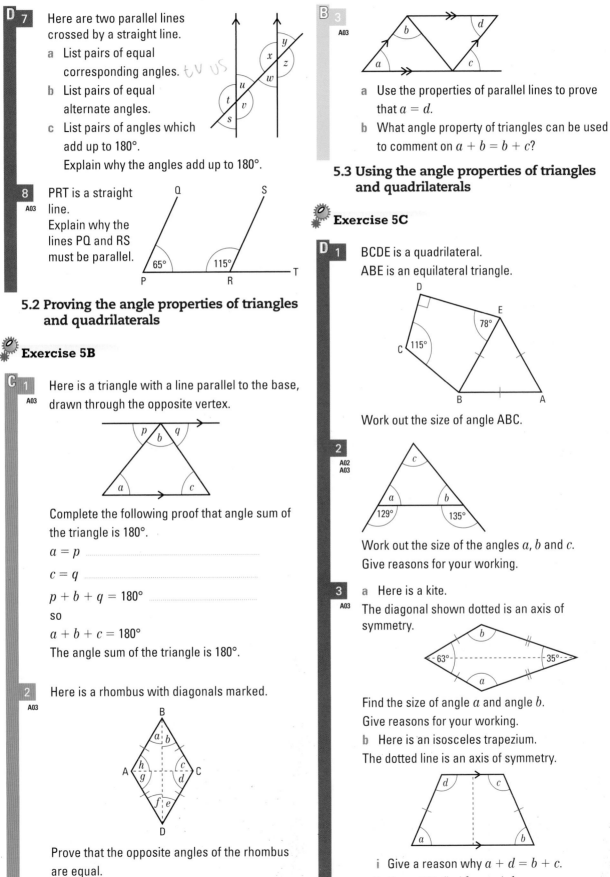

D 7 Here are two parallel lines crossed by a straight line.

a List pairs of equal corresponding angles. tv us

b List pairs of equal alternate angles.

c List pairs of angles which add up to 180°.
Explain why the angles add up to 180°.

8 PRT is a straight line.
A03
Explain why the lines PQ and RS must be parallel.

5.2 Proving the angle properties of triangles and quadrilaterals

Exercise 5B

C 1 Here is a triangle with a line parallel to the base, drawn through the opposite vertex.
A03

Complete the following proof that angle sum of the triangle is 180°.

$a = p$ ....

$c = q$ ....

$p + b + q = 180°$ ....

so

$a + b + c = 180°$

The angle sum of the triangle is 180°.

2 Here is a rhombus with diagonals marked.
A03

Prove that the opposite angles of the rhombus are equal.

B 3
A03

a Use the properties of parallel lines to prove that $a = d$.

b What angle property of triangles can be used to comment on $a + b = b + c$?

5.3 Using the angle properties of triangles and quadrilaterals

Exercise 5C

D 1 BCDE is a quadrilateral.
ABE is an equilateral triangle.

Work out the size of angle ABC.

2
A02
A03

Work out the size of the angles a, b and c.
Give reasons for your working.

3 a Here is a kite.
A03
The diagonal shown dotted is an axis of symmetry.

Find the size of angle a and angle b.
Give reasons for your working.

b Here is an isosceles trapezium.
The dotted line is an axis of symmetry.

i Give a reason why $a + d = b + c$.

ii If $a = 72°$, find b, c and d.

D 4
A03

Here is a quadrilateral.

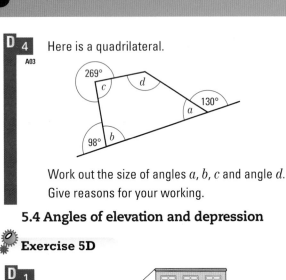

Work out the size of angles a, b, c and angle d.
Give reasons for your working.

5.4 Angles of elevation and depression

Exercise 5D

D 1
A03

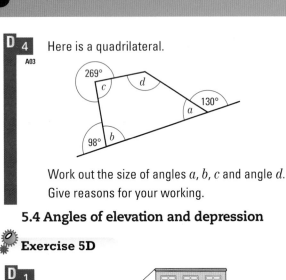

a What is the angle of depression of A from the top of the building?

b i Why does this angle equal a?

  ii What is angle a called?

2 The angle of depression of a point A on horizontal ground from the top of a flagpole is 35°.

a Show this information in a sketch.

b On your sketch show and label the angle of elevation of the top of the flagpole from A.

5.5 Bearings

Exercise 5E

1 In each of the following, give the bearing of B from A.

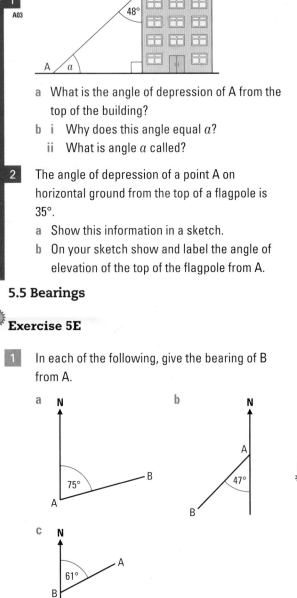

2 The diagram shows three towns A, B and C.

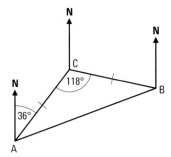

The bearing of C from A is 036°.
Angle ACB = 118°
CA = CB
Work out the bearing of

a B from A

b A from C

c B from C

D 3 The bearing of Leeds from Manchester is 037°.
Work out the bearing of Manchester from Leeds.

4
A02

A helicopter returns from York to its home base on a bearing of 148°.
Work out the bearing on which the helicopter flew from its home base to York.

C 5
A02
A03

This diagram shows the position of three telephone masts on a farmer's land.

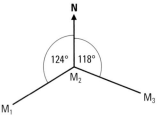

Find the bearing of:

a M_3 from M_2

b M_1 from M_2.

5.6 Using angle properties to solve problems

Exercise 5F

D 1
A02
A03

The diagram shows a parallelogram.

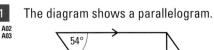

Work out the size of angle a.

C 2   L, M and N are points, as shown, on the sides of triangle ABC.

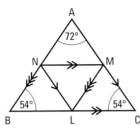

ML and AB are parallel.
NL and AC are parallel.
NM and BC are parallel.
Angle BAC = 72°
Angle ABC = 54°
Work out the size of each angle of triangle LMN.

3
A02
A03

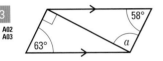

Work out the size of the angle *a*.
Give reasons for your working.

4
A02
A03
ABC is a straight line.

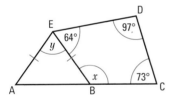

Work out the size of:

a   angle *x*

b   angle *y*.

Give reasons for your working.

5
A02
A03
In this quadrilateral $a + c = 180°$ and $b + d = 180°$.

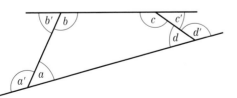

Prove that $a' + c' = 180°$ and $b' + d' = 180°$.

5.7 Polygons

Exercise 5G

D 1   For each polygon **a** to **e**, work out:

    i   the number of diagonals that can be drawn from one vertex

   ii   the number of triangles formed by the diagonals

  iii   the sum of the interior angles.

   a   quadrilateral

   b   heptagon

   c   nonagon

   d   dodecagon (12-sided polygon)

   e   *n*-sided polygon

2
A03
A square and a rhombus both have sides of equal length. Explain why a square is a regular polygon, but a rhombus, in general, is not.

Exercise 5H

D 1   Sam divides a regular polygon into 18 triangles
A02
A03
by drawing all the diagonals from one vertex.

   a   How many diagonals does Sam draw?

   b   How many sides does the polygon have?

   c   What is the size of each of the interior angles of the polygon?

2
A02
A03
Work out the size of each interior angle of:

   a   a regular nonagon

   b   a regular polygon with 15 sides.

C 3
A03
  a   Here is a pentagon. The sides of the pentagon are all the same length.

Work out the size of the angle *x*.
Give reasons for your answer.

   b   Two of these pentagons are placed as shown in the diagram.

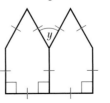

Work out the size of angle *y*.
Give reasons for your answer.

C **4**
A02
A03
Explain why the angle at the centre of a polygon can be a multiple of 9.

5
Here is an octagon.

a Work out the size of each of the angles marked with a letter.

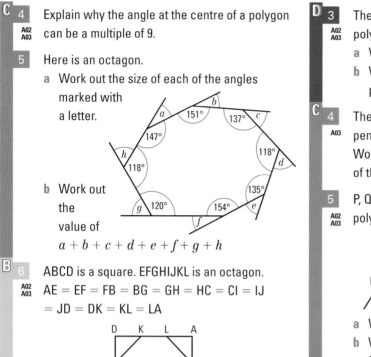

b Work out the value of $a + b + c + d + e + f + g + h$

B **6**
A02
A03
ABCD is a square. EFGHIJKL is an octagon.
AE = EF = FB = BG = GH = HC = CI = IJ
= JD = DK = KL = LA

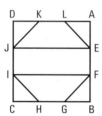

a What kind of triangles are AEL and JDK? What does this say about the exterior angles of the octagon at L and K?

b **i** ELKJ is a trapezium. What kind of trapezium is it? Give your reasons.

ii Work out the interior angle of the octagon at J. Does this mean that the octagon is regular? Give your reasons.

Exercise 5I

D **1**
One vertex of a polygon is the point X.

a Work out the size of the interior angle at X when the exterior angle at X is:
i 69° **ii** 38°.

b Work out the size of the exterior angle at X when the interior angle at X is:
i 132° **ii** 146°.

2
Work out the size of each exterior angle of:
a a regular hexagon
b a regular decagon
c a regular polygon with 15 sides
d a regular 20-sided polygon.

D **3**
A02
A03
The size of each exterior angle of a regular polygon is 24°.

a Work out how many sides the polygon has.

b What is the sum of the interior angles of this polygon?

C **4**
A03
The sizes of four of the exterior angles of a pentagon are 63°, 86°, 120° and 77°.
Work out the size of each of the interior angles of the pentagon.

5
A02
A03
P, Q and R are three vertices of a regular polygon. The exterior angle of the polygon is 20°.

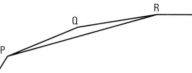

a Work out how many sides this polygon has.

b Work out the size of ∠QPR.
Give your reasons.

6 *
A02
A03
The diagram shows three sides, AB, BC, CD, of a regular polygon with centre O.

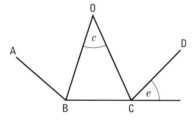

The angle at the centre of the polygon is c.
The exterior angle at vertex C is e.

a What type of triangle is BOC? Give reasons for your answer.

b Explain how you can find the sizes of angles c and e.

6 Collecting and recording data

Key Points

- **statistics:** an area of mathematics concerned with collecting and interpreting data.
- **statistical problem-solving process:**

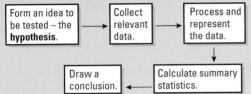

- **data:** the information that has been collected.
  - **primary:** data that has been collected by the person who is going to use it
  - **secondary:** data that has been collected by somebody else
  - **qualitative:** data described in words
  - **quantitative:** data given as a numerical value
  - **discrete:** quantitative data that can only take certain numerical values
  - **continuous:** quantitative data that can take any numerical value
- **population:** the whole group of people you wish to find something out about.
- **sample:** a select number of people from the population that can be surveyed to represent to total population.
- **a biased sample:** an unfair sample that may occur when:
  - the sample does not truly represent the population
  - the sample is too small
- **random sample:** a sample that has given every member of the population an equal chance of being chosen.

- **strata:** non-overlapping groups within a population that share similar characteristics.
- **stratified sample:** a sample in which the population is split into strata, and a simple random sample is taken from each stratum.
- **frequency:** the total number of times a certain event happens.
- **class interval:** a group of numbers that data can fall into. Often used when recording continuous data that is widely spread.
- **class limits:** the upper and lower bounds of the class interval.
- **class size:** the difference between the upper and lower bounds of the class interval.
- **questionnaire:** a list of questions designed to collect data.
- **open question:** one that has no suggested answers.
- **closed question:** one that has a set of answers to choose from.
- **two-way table:** a table that shows how data falls into two different categories.
- **database:** an organised collection of information from which secondary data can be obtained.
- **finding the number to be selected from a stratum:** use the formula
  The number sampled in a stratum =
  $$\frac{\text{number in stratum}}{\text{number in population}} \times \text{total sample size}$$
- **collecting data by observation:** use a data collection sheet to record a tally of how often a certain event happens.

6.1 Introduction to statistics

Exercise 6A

Questions in this chapter are targeted at the grades indicated.

1 Write down whether each of the following is secondary or primary data.
   a Data collected by you from a magazine.
   b Data collected by you from a weather website.
   c Data collected by you questioning fellow students at your school.

2 Write down whether the following are qualitative or quantitative data.
   a The sizes of shoes worn by a group of people.
   b The heights of members of a cricket team.
   c The colours of cars in a car park.
   d The areas of walls in an art gallery.

3 Write down whether the following are continuous or discrete data.

a The number of plants in a greenhouse.

b The number of people at a pop concert.

c The weight of apples.

d The time taken to swim 400 m.

D 4 A car designer wants to find out how comfortable the passenger seats are in a new model GT.

A03

a Write down a hypothesis she could use.

b What is the next thing that she should do?

6.2 Sampling methods

Exercise 6B

C 1 One way to generate random numbers is to pick them out of a hat.

Give two other ways in which you can generate random numbers.

2 Explain why a sample may need to be used.

B 3 An office department has 55 members of staff. Six are to be selected to train as fire marshals. The manager decides to choose a simple random sample of six.

She uses a computer to generate random numbers. These are the first few numbers she generates.

2 0 3 3 3 1 5 3 4 6 3 3 6 2 7 1 4 6 8 4 0 6 …

Describe how she could use these numbers to get her sample of the department's staff.

6.3 Stratified sampling

Exercise 6C

B 1 A head of Year 9 wants to find out what his students think

ResultsPlus
Exam Tip

Always make sure your individual samples total the required sample size.

about the school's method of choosing subjects for GCSE. He decides to ask a stratified sample of 60 students. The table shows the total number of boys and the total number of girls in Year 9.

Boys	Girls
252	288

Work out the number of boys and the number of girls he should include in the sample.

A 2 * A canoe club allows members to tackle different rapids according to their level of experience. Those who have passed the beginners' course are only allowed out on the easiest rapids. Those who have passed the advanced course are allowed on the more difficult rapids. There are 120 members at beginner level and there are 300 advanced members. Describe exactly how you would find a stratified sample of 10% of the club's members.

A03

3 * A hospital manager wants to find out what her employees think about the catering facilities at the hospital. She decides to ask a stratified sample of 80 of her employees. The table shows how many people are in each of the six strata she intends to use.

A02

	Office staff	Domestic staff	Medical staff
Females	100	60	220
Males	160	140	280

Calculate the number of employees she needs to ask in each strata and describe how she should pick the individual members of each strata.

6.4 Collecting data by observation and experiment

Exercise 6D

1 The controller of a bus station keeps a record of the vehicles arriving at the station during a five-minute interval. This data is listed below:

Single-decker	Taxi	Minibus
Coach	Double-decker	Double-decker
Single-decker	Double-decker	Double-decker
Minibus	Single-decker	Single-decker
Double-decker	Taxi	Coach
Double-decker	Taxi	Taxi
Taxi	Single-decker	Single-decker
Coach	Single-decker	Minibus
Coach	Double-decker	Coach
Single-decker	Single-decker	Coach

a Draw a tally chart to show this data.

b Write down the name of the least common type of vehicle.

c Write down the name of the most common type of vehicle.

D 2
A02

A radio presenter asks listeners to text in the number of CDs they have bought in the past three months.
He receives the following data.

6	1	2	3	6	6
6	7	1	0	3	6
1	4	3	2	8	8
6	5	8	10	8	4
4	5	5	3	7	6
7	4	3	5	4	3
6	6	7	8		

> **Results Plus**
> **Watch Out!**
> Make sure classes don't overlap.

Draw and fill in a data collection sheet showing this information. Use equal class intervals starting with the class 0–2.

3
A fruit farmer weighs 30 apples produced from his orchard.
The weights, in grams, are shown below:

111	110.5	125	124	123
115	112	120	115	123
116.3	133	130	117	121
129.7	134	113	129	128
115	116	126	127	120
124	130	121	115	110

Weight (w)	Tally	Frequency
$110 \leqslant w < 115$		
$115 \leqslant w < 120$		
$120 \leqslant w < 125$		
$125 \leqslant w < 130$		
$130 \leqslant w$		

a Copy and complete the data collection sheet for this data.

b Write down the most common class.

c Write down the least common class.

6.5 Questionnaires

Exercise 6E

D 1
A03

A questionnaire includes the following question.
'Dogs should always be walked on a lead.
Do you agree?'

Yes ☐   No ☐

What is wrong with this question?

D 2

A town council wants to find out whether residents would like a new sports centre to be built in the town. The council decides to use a questionnaire. The following questions are suggested.

A: What is your opinion of sports centres?

B: Does the town need a new sports centre?
   Yes/No

C: Would you use a sports centre?
   Yes/No

D: Can you suggest a site for a sports centre?

Which of the above are open questions and which are closed?

C 3
A03

Another town council asks parents if changes should be made to the playground in the local park. They design a questionnaire to find out what changes the parents would like. The following questions are suggested.
Write down what is wrong with each of them and design a new question for each that is more suitable.

a What do you think of the present facilities?

Very good ☐   Good ☐   Satisfactory ☐

b What would you like to see in the playground?

Slide ☐   Swing ☐   Sandpit ☐

c How often do you take your children to the park?

Often ☐   Not often ☐

4 *
A03

A restaurant has launched a new menu and the manager wants to find out if guests like it.
She decides to use a questionnaire.
Write down a suitable question she could use.

6.6 Two-way tables

Exercise 6F

> **Results Plus**
> **Exam Tip**
> When completing two-way tables, look for rows with only one number missing and fill these in first.
> The numbers in each row must add up to the row total and the same goes for columns.

C 1
A02
A number of workers in a factory were asked which type of drink they wanted to be supplied in the drinks machine. A total of 16 workers wanted tea, of which 7 were men.

8 women wanted hot chocolate. 10 men and 12 women wanted coffee. There were 25 men in total.

a Draw and complete a table of the data.

b How many wanted hot chocolate?

c How many workers were asked in total?

2 In a restaurant survey 25 men and 25 women were asked whether they preferred pizza or pasta. 10 men preferred pizza. 12 women preferred pasta.

a Draw up a two-way table to show this information.

b How many people liked pasta best?

3 Lewis's music collection contains only hip-hop and rock albums. He has 14 albums on CD, of which 4 are rock. He has 12 albums on vinyl, of which 4 are hip-hop. He also has 130 MP3 albums, of which 35 are rock.

a Draw up a two-way table to show this information.

b Write down the number of rock albums in Lewis's collection.

c Write down the total number of albums in Lewis's collection.

6.7 Sources of bias

Exercise 6G

C 1 A university wants to get information on sixth form colleges' views regarding how they respond to queries. They send a questionnaire to a sample of colleges in the Birmingham area. Is this a biased sample?
Give one reason for your answer.

2 Write down, with reasons, whether or not each of the following are biased.

A: A manufacturer of electric toothbrushes wants to test whether the toothbrushes work as advertised. He tests a sample, using every twelfth toothbrush made.

C B: A group of doctors want to know how often people use their practice. They ask all the people attending on one particular Tuesday.

C: You ask 60 people using a local swimming pool what they think about its facilities.

D: A market research firm wants to find out which utility company people prefer. They do a telephone poll of 25 people in each of 12 towns.

**3 * **
A03
Two hundred and fifty parents march to the local town hall to demand that a mobile telephone mast is not built next to a school.

Gary says 'If so many people march, they must be right.'

Nell says 'I disagree.'

Discuss the views of Gary and Nell.

6.8 Secondary data

Exercise 6H

1 The following database gives some information about the population of 5 cities.

	Year					
	1980	1985	1990	1995	2000	2005
Population of City A (000s)	294	313	325	326	366	391
Population of City B (000s)	4745	4745	4790	4853	4857	4949
Population of City C (000s)	2981	2990	3013	3001	3034	3054
Population of City D (000s)	391	402	439	459	462	490
Population of City E (000s)	532	531	550	559	572	583

a What was the population of City C in 1990?

b Which city has the highest population?

c In which year was the population of City E at its lowest?

d Which city's population has increased the least from 1980 to 2005?

2 The following database gives information about the weather in a seaside resort during the last six months of the year.

	Max temp (°C)	Min temp (°C)	Sea temp (°C)	Total sunshine (hours)	Total rainfall (mm)	Number of rainy days
July	21.0	14.5	18.1	237.7	40.9	11
August	20.0	15.4	18.4	154.2	81.5	21
September	18.3	12.0	16.9	161.0	65.6	12
October	15.0	8.4	13.8	145.7	69.7	17
November	11.6	6.8	11.1	47.2	134.4	21
December	7.9	3.1	7.7	87.3	29.2	11

a How many days of rain were there in October?
b Write down the month that had the most sunshine.
c In which month was the sea the coldest?
d Which month had the least difference between maximum and minimum temperatures?

C 3 The database gives the ski conditions at various resorts.

Resort	Snow depth (cm): lower slopes	Snow depth (cm): upper slopes	Temperature (°C)
Aspen (USA)	64	132	−14
Avoriaz	100	130	1
Cervinia	90	190	−3
Cortina	30	120	−1
Courchevel	50	80	−1
Davos	29	110	−1
Kitzbuhel	25	55	3
La Plagne	70	135	−2
Les Arcs	30	70	−2
Mayrhofen	5	60	2
Meribel	29	107	−1
Soldeu	30	40	0
Tignes	68	110	−3
Vail	51	51	−15
Val d'Isere	68	110	−3
Val Thorens	80	110	−3
Verbier	35	140	−2
Wengen	20	60	0
Whistler (Canada)	239	239	−7
Zermatt	40	146	−5

a What is the difference between the lowest temperature and the highest temperature?
b What do you notice about the depth of snow on the lower slopes of Cortina, Les Arcs and Soldeu?
c List any resorts with 100 cm or more difference between upper and lower slope depths.
d Comment on the depth of snow on both upper and lower slopes at Vail and Whistler. Suggest a reason for what you notice.

A03

7 Measure

🔧 Key Points

- **compound measure:** a measure which involves two units such as km per hour. Often a measure of a rate of change.
- **average speed, time and distance formula:**
  if S = average speed, D = total distance travelled and T = total time taken
  - $S = \dfrac{D}{T}$
  - $T = \dfrac{D}{S}$
  - $D = S \times T$

  - speed is a compound measure usually measured in miles per hour (mph), kilometres per hour (km/h), metres per second (m/s)
- **density, mass and volume formula:** if D = density, M = mass and V = volume
  - $D = \dfrac{M}{V}$
  - $V = \dfrac{M}{D}$
  - $M = D \times V$

  - density is a compound measure usually measured in kg per m³ (kg/m³) or g per cm³ (g/cm³)

- **converting metric units:** when changing from a large unit to a smaller unit, multiply by 10, 100 or 1000. When changing from a small unit to a larger unit, divide by 10, 100 or 1000.

Length	Weight	Capacity/Volume
10 mm = 1 cm	1000 mg = 1 g	100 cl = 1 litre
100 cm = 1 m	1000 g = 1 kg	1000 ml = 1 litre
1000 mm = 1 m	1000 kg = 1 tonne	1000 cm³ = 1 litre
1000 m = 1 km		1000 l = 1 m³

- **converting between metric and imperial units:** use the facts in this table:

Metric unit	Imperial unit
1 kg	2.2 pounds
1 litre (l)	$1\frac{3}{4}$ pints = 1.75 pints
4.5 l	1 gallon
8 km	5 miles
30 cm	1 foot
2.54 cm	1 inch

7.1 Converting between units of measure

⚙ Exercise 7A

Questions in this chapter are targeted at the grades indicated.

1 Convert these lengths to centimetres.
   a 8 m   b 340 mm   c 6.7 m
   d 0.78 m   e 63 mm   f 238 mm
   g 4 km   h 0.018 km

2 Convert these weights to kilograms.
   a 4 tonnes   b 6.5 tonnes   c 2000 g
   d 700 g   e 580 g   f 3900 g

3 Convert these volumes to litres.
   a 3000 ml b 800 cl c 6200 ml
   d 27 000 ml

4 Wooden crates have a height of 80 cm. How many metres high is a stack of 5 of these crates?

⚙ Exercise 7B

1 Convert 3 kg to pounds.

2 Convert 220 pounds to kilograms.

3 Convert 14 pints to litres.

4 Convert 45 litres to gallons.

5 Convert 40 litres to pints.

6 Convert 10 feet to centimetres.

7 Convert 80 km to miles.

8 Convert 50 miles to kilometres.

9
A02 A03 One litre of pineapple juice costs 114p. Ben needs a gallon of juice to make fruit punch for a party. How much will this cost?

ResultsPlus
Watch Out!

The values in the metric and imperial conversion table given in the Key Points are not exact; they are the rough equivalents that need to be used in examinations.

Exercise 7C

1. There are 8 pints in a gallon.
   a Convert 72 pints to gallons.
   b Convert 60 gallons to pints.

2. **A02** There are 28.3 grams in an ounce and there are 16 ounces in 1 pound.
   Work out how many grams there are in 1 pound.

3. **A02** There are 14 pounds in a stone. The soil in a skip weighs 8 stones and 9 pounds.
   a Work out the weight of the soil in pounds.
   b Work out the weight of the soil in kilograms.

7.2 Compound measures

Exercise 7D

Results Plus
Exam Tip

The units of a compound measure will tell you what to do, so km/*l* will mean distance ÷ volume.

C 1 **A03** A van travels between two towns, 250 km apart, and uses 18 litres of petrol.
   a Work out the average rate of petrol usage. Give the answer in km/litre.
   b Estimate the amount of petrol used by the time the driver stops for a break after 68 km.

2. **A02** A bubble machine is designed to blow bubbles 3 times every minute.
   a Work out how many times the machine blows bubbles in an hour.
   b The machine holds enough liquid to blow bubbles 720 times. How long will it take for the machine to run out of liquid?

3. **A02 A03** A factory produces 270 litres of ice cream every 5 minutes.
   a Work out the average rate of production. Give units with the answer.
   b The factory has an order of 2300 litres of chocolate brownie ice cream. How long will it take to fulfil the order? Give the answer in minutes and seconds.

4. **A03** A hybrid car averages 18 km per litre of petrol. How many litres of petrol does the car use per kilometre?

7.3 Speed

Exercise 7E

Results Plus
Watch Out!

It is important to be careful with time; it is best to use decimals and remember that there are 60 minutes in 1 hour.

D 1 Hannah competes in a 10 km run. She completes the race in $2\frac{1}{4}$ hours. What was her average speed?
   Give your answer correct to 3 significant figures.

C 2 Dinesh set off for work at 8 am. His office is 24 km away. He got stuck in a big traffic jam on the way there and didn't arrive until 11 am. Work out Dinesh's average speed.

3. A snow plough travels at an average speed of 12 km/h and covers 21 km. How long, in hours and minutes, does this take?

4. **A03** Convert 90 m/s into km/h.

5. **A02 A03** John drives from his home to visit a friend. For the first 3 hours he drives at an average speed of 40 km/h. He then drives the remaining 60 km to his friend's house at an average speed of 30 km/h.
   Work out the average speed for John's journey.

7.4 Density

Exercise 7F

C 1 75 cm³ of a substance has a mass of 1200 g. Work out the density of the substance.

2. Platinum has a density of 21.4 g/cm³. A brooch contains 16 g of platinum. Work out the volume of platinum in the brooch.

3. 98.6 g of magnesium has a volume of 58 cm³. Work out the density of magnesium.

B 4 **A02 A03** The density of copper is 8930 kg/m³; the density of silver is 10 490 kg/m³.
   A block of copper has a volume of 0.5 m³ and a block of silver has a volume of 0.3 m³.
   Which of the two blocks has the greater mass and by how much?

8 Congruence, symmetry and similarity

Key Points

- **congruent triangles:** two triangles that have exactly the same shape and size.
- **conditions of congruence:**
  - **SSS:** the three sides of each triangle are equal
  - **SAS:** two sides and the included angle are equal
  - **AAS:** two angles and a corresponding side are equal
  - **RHS:** each triangle contains a right angle, and the hypotenuses and another side are equal
- **line symmetry:** when a shape can be folded so that one part of the shape fits exactly on top of the other part. The dividing line is the line of symmetry or mirror line.
- **rotational symmetry:** occurs if a shape looks the same as is did at 0° during a 360° rotation. The number of times is the order of rotational symmetry.
- **quadrilateral:** a polygon with four sides.
- **types of quadrilaterals:**
  - **square:** 4 equal sides, 4 right angles

  - **rectangle:** 2 pairs of equal and parallel sides, 4 right angles

- **parallelogram:** 2 pairs of equal and parallel sides

- **trapezium:** 1 pair of parallel sides

- **isosceles trapezium:** 1 pair of parallel sides, other sides equal

- **kite:** 2 pairs of equal adjacent sides

- **rhombus:** 4 equal sides, 2 pairs of parallel sides

- **similar shapes:** when one shape is an enlargement of the other shape. Corresponding angles are equal and corresponding sides are in the same ratio.
- **similar triangles:** two triangles with any of the following:
  - corresponding angles are equal
  - corresponding sides are in the same ratio
  - one equal angle and adjacent sides are in the same ratio

8.1 Congruent triangles

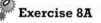 **Exercise 8A**

> Questions in this chapter are targeted at the grades indicated.

A 1 *
A03

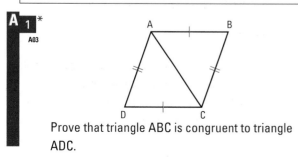

Prove that triangle ABC is congruent to triangle ADC.

A 2 *
A03
ABC is an isosceles triangle. D is the midpoint of BC.

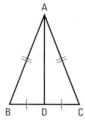

a Prove that triangles ABD and ACD are congruent.

b Hence prove that angle ADB is a right angle.

A **3** * PQRS is a square. PABC is a square.

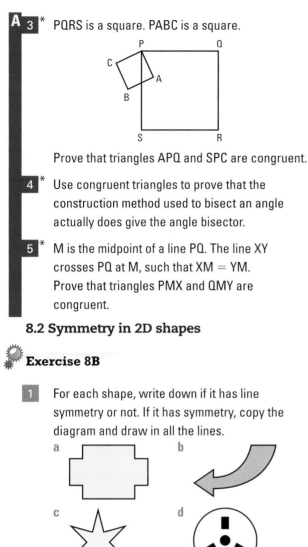

Prove that triangles APQ and SPC are congruent.

4 * Use congruent triangles to prove that the construction method used to bisect an angle actually does give the angle bisector.

5 * M is the midpoint of a line PQ. The line XY crosses PQ at M, such that XM = YM. Prove that triangles PMX and QMY are congruent.

8.2 Symmetry in 2D shapes

Exercise 8B

1 For each shape, write down if it has line symmetry or not. If it has symmetry, copy the diagram and draw in all the lines.

a   b

c   d

e   f

2 Using tracing paper if necessary, state which of the shapes in Question 1 have rotational symmetry and which do not have rotational symmetry. For the shapes that have rotational symmetry, write down the order of the rotational symmetry.

ResultsPlus
Watch Out!

Many shapes have the same number of lines of symmetry as their order of rotational symmetry, but this is not always so.
Don't confuse the two!

3

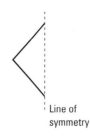

Line of symmetry

a Copy and complete this shape so that it has line symmetry.

b Write down the name of the complete shape.

4 Each diagram shows an incomplete pattern. For part **a**, copy the diagram and shade eight more squares so that both dotted lines are lines of symmetry of the complete pattern. For part **b**, shade seven more squares so that the complete pattern has rotational symmetry of order 4.

a

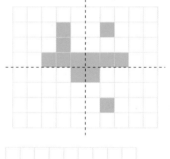

b

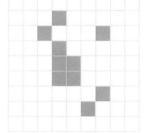

5 a Draw a shape that has three lines of symmetry and rotational symmetry of order 3.
A03
b Draw a shape with one line of symmetry and rotational symmetry of order 2.

c Draw a shape that has no lines of symmetry and rotational symmetry of order 3.

8.3 Symmetry of special shapes

Exercise 8C

1 a On squared paper, draw a rectangle. On your
A02
A03 rectangle, draw all the lines of symmetry.

b What can you say about the shapes formed by the lines of symmetry?

2 Gemma wants to draw a quadrilateral.

A03 Dean tells her to draw one with four equal sides and opposite sides parallel.

a Gemma says that there are two quadrilaterals that fit Dean's description. What are they?

b Dean tells her that this one has opposite angles that are equal. Can Gemma draw it now? Explain your answer.

c Dean then says that this one has two lines of symmetry and rotational symmetry of order 2. What quadrilateral is he thinking of?

3 Draw a non-regular polygon which has two lines
A02
A03 of symmetry.

4 Draw a non-regular polygon which has rotational
A02
A03 symmetry of order greater than 2.
State the order of rotational symmetry of your polygon.

8.4 Recognising similar shapes

Exercise 8D

D 1 State which of the pairs of shapes are similar.

a

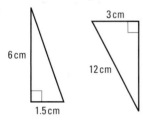

b

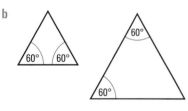

B 2 Show that hexagon ABCDEF is similar to hexagon GHIJKL.

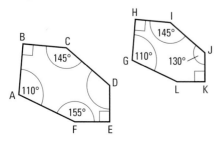

Exercise 8E

B 1 Two tubes of toothpaste are similar. The travel
A02
A03 size has a width of 2.5 cm and a length of 8.4 cm. The normal size has a width of 3.2 cm. What is its length? Give your answer to three significant figures.

2 The diagram shows the corresponding parts of
A02
A03 two similar model boats.

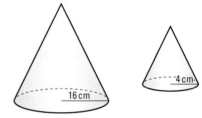

Calculate the value of:

a x

b y.

A 3 These cones are similar. The height of the smaller cone is 9 cm.

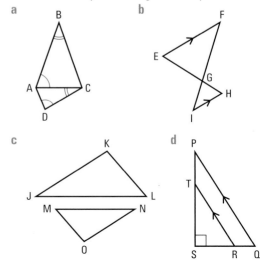

Find the height of the larger cone.

8.5 Similar triangles

Exercise 8F

C 1 For each pair of similar triangles:

i name the three pairs of corresponding sides

ii state which pairs of angles are equal.

a b

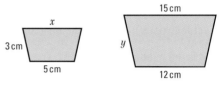

c d

A 2
A03

Triangle PQR is similar to triangle STU.

∠PQR = ∠STU

PQ = 4 cm, PR = 3.5 cm, ST = 2 cm, TU = 1.6 cm

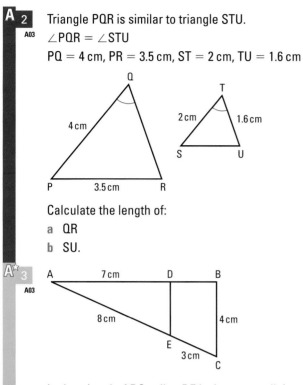

Calculate the length of:

a QR

b SU.

A★ 3
A03

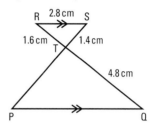

In the triangle ABC, a line DE is drawn parallel to BC.

AD = 7 cm, AE = 8 cm, BC = 4 cm, EC = 3 cm

a Prove that triangle ADE is similar to triangle ABC.

b Calculate the length of DE.

c Calculate the length of DB.

A★ 4
A03

The diagram shows triangle PQR which has a line ST drawn across it. ∠RPS = ∠TSQ

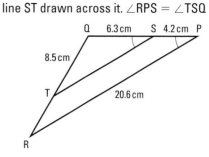

a Prove that triangle PQR is similar to triangle SQT.

b Calculate the length of ST.

c Calculate the length of QR.

Give your answers correct to three significant figures.

5
A03

In the diagram PQ is parallel to RS.

R 2.8 cm S
1.6 cm T 1.4 cm
4.8 cm
P Q

a Prove that triangle RST is similar to triangle PQT.

b Calculate the lengths of:

   i PQ   ii PT.

9 Expanding brackets and factorising

Key Points

- **expanding brackets:** multiply each term inside the bracket by the term outside the bracket.
- **factorising:** the opposite of expanding brackets. Find a common factor of the terms, write it outside the brackets, then decide what is needed inside the brackets.
- **multiplying two brackets:** multiply each term in the first bracket by the second bracket, expand the brackets and simplify the resulting expression. Alternatively use the grid method.

- **factorising the quadratic expression** $x^2 + bx + c$:
  - find a pair of numbers whose product is $+c$ and sum is $+b$
  - use these two numbers, p and q, to write down the factorised from $(x + p)(x + q)$
- **using the difference of two squares:** any expression that can be written in the form $a^2 - b^2$ can be factorised using the result $a^2 - b^2 = (a + b)(a - b)$.

9.1 Expanding brackets

Exercise 9A

Questions in this chapter are targeted at the grades indicated.

D **1** Expand

a $3(x + 4)$ b $2(z - 2)$

c $3(m + n)$ d $4(6 - y)$

e $3(x + 2y - 4)$ f $4(2d + 5)$

g $3(x^2 + 2)$ h $2(g^2 - 3g + 2)$

C **2** Expand

a $p(p + 3)$ b $q(q - 2)$

c $2x(x + 4)$ d $h(3 - h)$

e $c(b + a)$ f $s(6s - 5)$

g $2t(4t + 2)$ h $5x^2(x - 7)$

3 Expand

a $-2(b + 4)$ b $-2(2x + 7)$

c $-m(m + 3)$ d $-4y(5y + 4)$

e $-4(h - 2)$ f $-4g(2 - g)$

g $-2z(z - 4)$ h $-2n(4m + 5n - 6)$

Exercise 9B

ResultsPlus
Watch Out!

You must multiply out the brackets before you collect like terms.
Check your signs.

C **1** Expand and simplify

a $5(z - 1) + 3z$ b $4p + 2(p + 1)$

c $4(h + 1) + 3h$ d $2(d + 3) + 4(d - 7)$

e $6a + b + 4(a + b)$

f $4(5m - n) + 5(n - m + 3)$

2 Expand and simplify

a $4(y + 9) - 2(y + 7)$

b $5(2a + 1) - 2(a + 6)$

c $k - 4(k + 3)$

d $d(d + 3) - 2(d + 1)$

e $3n(n - 4) - n(5n + 1)$

f $4t(2 + 5t) - 3t(1 + t)$

3 Expand and simplify

a $4(t - 4) - 5(t - 1)$

b $4(h + 3) - 2(h - 6)$

c $3g(g + 2) - g(g + 2)$

d $5e(2e - 3) - e(4 - e)$

e $3s(s + 5) - 2(2 - s)$

f $q(p + q) - p(p - q)$

4 Expand and simplify

a $9s - 5(s + 1)$ b $12r + 4(r + 7)$

c $6f^2 - 3f(f + 2)$ d $8n + 2n(n - 1)$

e $2g - 3g(g - h)$ f $5p - 2p(2 - p)$

9.2 Factorising by taking out common factors

Exercise 9C

D 1 Factorise

a $10x + 5$ b $2y - 8$ c $2y + 10z$

d $21g - 7$ e $10s + 2t$ f $6j + 18k$

g $45u + 5v + 20w$

h $ac - bc$ i $u - uv$ j $6x^2 + 8x + 2$

k $4h^2 - 3h$ l $4q^2 - 7q$ m $4y^2 + y$

n $3b - 5b^2$ o $q^3 + 6q$ p $a^2 + 2a^3$

C 2 Factorise completely

a $3xy + 3xt$ b $2ab - 6ac$

c $8pq + 4ps$ d $12xy - 4y$

e $4ab + 12ac + 8ad$ f $mpq - mq$

g $6x^2 + 3x$ h $12t^2 - 36t$

i $6f^2 + 12f^3$ j $3h^4 + h^2$

k $4cd^2 - 16c^2d$ l $2a^3b + 2ab^3$

m $8pqr + 16prs$ n $14a^2b - 8ab^2 + 20ab$

o $10x^2y - 25x^2y^2$ p $(2g)^2 + 2g$

Exercise 9D

B 1 Factorise

a $(x + 4)^2 + 2(x + 4)$

b $x(x + y) + y(x + y)$

c $z(z + 4) + 5z$

d $(5a + b)(3a - b) + 3(3a - b)$

e $(a - 9)^2 + 2(a - 9)$

f $(3g + 2)^2 - 2(3g + 2)$

2 Factorise completely

a $12(h + 2)^2 + 4(h + 2)$

b $25(x - 3)^2 - 10(x - 3)$

c $8(c + 5)^2 + 16(c + 5)$

d $3(q + 4) + 6(q + 4)^2$

e $6(a + b)(a - b) - 24(a + b)$

f $8x^2(x - 2) + 6x(x - 2)$

9.3 Expanding the product of two brackets

Exercise 9E

B 1 Expand and simplify

a $(x + 2)(x + 4)$ b $(x + 3)(x + 2)$

c $(x + 3)(x - 5)$ d $(y - 3)(y + 2)$

e $(y + 2)(y - 1)$ f $(z - 4)(z - 6)$

g $(g - 7)(g - 3)$

B

h $(x + 3)^2$

i $(d + 6)^2$

j $(p - 5)^2$

k $(x - y)^2$

l $(p + q)^2$

ResultsPlus
Watch Out!

Note that $(a + b)^2$ is not equal to $a^2 + b^2$.

A 2 Expand and simplify

a $(x + 3)(2x + 3)$ b $(x - 1)(2x + 1)$

c $(2x + 4)(x + 3)$ d $(y - 2)(3y + 2)$

e $(2z + 3)(z + 1)$ f $(2t + 3)(3t + 1)$

g $(3k + 12)(2k + 3)$ h $(2a - 3)(2a + 4)$

i $(3y + 1)(4y - 3)$ j $(2s - 2)(3s - 1)$

k $(3p + 1)^2$ l $(2d - 3)^2$

3 Expand and simplify

a $(x + 3y)(x + 2y)$ b $(x - 3y)(x + 2y)$

c $(x + 3y)(x - 2y)$ d $(x - 3y)(x - 2y)$

e $(2a + 3b)(3a - 2b)$ f $(3m - 2n)(2m - 3n)$

g $(4x + 5y)^2$ h $(4x - 5y)^2$

9.4 Factorising quadratic expressions

Exercise 9F

1 Write down a pair of numbers:

a whose product is $+12$ and whose sum is $+7$

b whose product is $+21$ and whose sum is -10

c whose product is $+15$ and whose sum is -8

d whose product is -6 and whose sum is $+1$

e whose product is -12 and whose sum is -4

f whose product is -16 and whose sum is 0.

A 2 Factorise

a $x^2 + 5x + 6$ b $x^2 + 9x + 20$

c $x^2 + 13x + 42$ d $x^2 - 5x + 6$

e $x^2 - 9x + 20$ f $x^2 - 13x + 42$

g $x^2 - x - 6$ h $x^2 - x - 20$

i $x^2 - x - 42$ j $x^2 + 4x - 5$

k $x^2 + 6x - 112$ l $x^2 - 25$

m $x^2 - 100$

Exercise 9G

A 1 Factorise

a $x^2 - 81$

b $x^2 - 64$

c $y^2 - 121$

d $49 - y^2$ e $z^2 - 900$ f $225 - z^2$

g $(x + 1)^2 - 5$ h $64 - (8 - y)^2$

i $(c - d)^2 - (c + d)^2$

ResultsPlus
Exam Tip

It will help you in the examination if you learn $a^2 - b^2 = (a + b)(a - b)$.

A 2

A02

Without using a calculator, use algebra to find the value of:

a $64^2 - 49^2$ b $4.5^2 - 3.5^2$

c $0.375^2 - 0.125^2$ d $905^2 - 895^2$

3 Factorise these expressions, simplifying your answers where possible.

a $9x^2 - 100$ b $4a^2 - 1$

c $8t^2 - 144$ d $16 - (z + 2)^2$

e $(2g - 1)^2 - (2g + 1)^2$

f $(a + b - 1)^2 - (a + b + 1)^2$

g $36(a + \frac{1}{2})^2 - 144(b - \frac{1}{2})^2$

h $16(m + n)^2 - 16(m - n)^2$

4 Factorise completely

a $3x^2 - 27$ b $5h^2 - 20$

c $40r^2 - 9000$ d $4p^2 - 36q^2$

e $12u^2 - 48v^2$

f $3(d + 1)^2 - 3(d - 1)^2$

Exercise 9H

A 1 Factorise

a $5x^2 + 8x + 3$ b $2x^2 + 7x + 5$

c $3x^2 + 7x + 4$ d $6x^2 - 9x - 6$

e $6x^2 + 10x - 5$ f $6x^2 + 8x - 3$

g $5x^2 + 23x + 12$ h $4x^2 - 7x + 3$

i $8y^2 + 10y - 3$ j $2y^2 - 13y + 21$

k $7y^2 - 27y - 4$ l $3y^2 - 3y - 6$

m $4z^2 - 8z - 5$ n $6z^2 - 7z - 3$

o $6z^2 - 22z + 12$

2 Factorise completely

a $6x^2 - 13x + 5$ b $6y^2 - 25y + 4$

c $5x^2 - 7x - 6$

A☆ 3 Factorise

a $x^2 + 2xy - 3y^2$ b $2x^2 + 11xy + 5y^2$

c $6x^2 + 3xy - 3y^2$

10 Area and volume 1

> ### Key Points

- **area:** the amount of space inside a 2D shape.
  - **area of a square** = length × length = l^2
  - **area of a rectangle** = length × width = $l \times w = lw$
  - **area of a triangle** = $\frac{1}{2}$ × base × vertical height = $\frac{1}{2} \times b \times h = \frac{1}{2}bh$
  - **area of a parallelogram** = base × vertical height = $b \times h = bh$
  - **area of a trapezium** = $\frac{1}{2}$ × sum of parallel sides × distance between them = $\frac{1}{2}(a + b)h$
  - measured in square units, such as square millimetres (mm²), square centimetres (cm²), square metres (m²), square kilometres (km²)
- **perimeter:** the total distance around the edge of a 2D shape.
- **circle formula:** if C = the circumference and d = the diameter
  - $\frac{C}{d} = \pi$
  - $C = \pi d = 2\pi r$
  - $d = \frac{C}{\pi}$

- **area of a circle (A):**
  $A = \pi r^2$
- **net:** a 2D shape that can be folded into a 3D shape.
- **plans and elevations:** 2D views of a 3D object drawn from different angles.
  - **plan:** the view from above
  - **front elevation:** the view from the front
  - **side elevation:** the view from the side
- **volume:** the amount of space a 3D shape takes up.
  - volume of a prism = the area of the cross section × length
  - volume of a cylinder = $\pi r^2 h$
- **finding the area of compound shapes:** split the shape into a number of simpler shapes. The total area = the sum of the areas of each part.
- **drawing 3D objects:** use isometric paper with the vertical lines going down the page and no horizontal lines.

10.1 Area of triangles, parallelograms and trapeziums

Exercise 10A

Questions in this chapter are targeted at the grades indicated.

D 1 Work out the areas of these triangles and parallelograms.

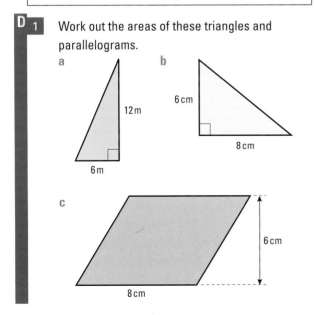

a 12 m, 6 m

b 6 cm, 8 cm

c 6 cm, 8 cm

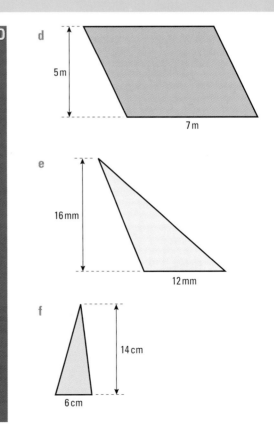

D d 5 m, 7 m

e 16 mm, 12 mm

f 14 cm, 6 cm

D **2** Copy and complete this table.

Shape	Base	Height	Area
Triangle	12 cm	7 cm	
Triangle	8 cm	18 cm	
Triangle	9 cm		54 cm²
Parallelogram	11 cm	7 cm	
Parallelogram		6 cm	78 cm²

3
A03
a A rectangle has a length of 6 m and an area of 54 m². Work out the width of the rectangle.

b A square has an area of 225 cm².
Work out the length of the sides of the square.

Exercise 10B

C **1** Work out the area of each of these trapeziums.

a

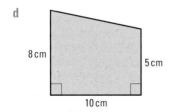

b

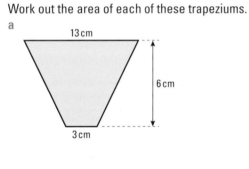

c

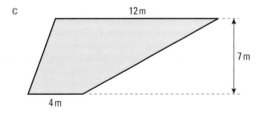

d

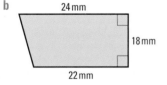

10.2 Problems involving perimeter and area

Exercise 10C

D **1**
A02
This floor is to be carpeted. Gripper rods are to be fitted round the perimeter of the room.

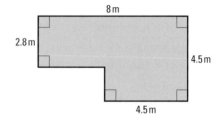

a Work out the length of rods needed.
Give the units of your answer.

b Work out the area of carpet required.
Give the units of your answer.

2
A02
A03
Fiona wants to lay a wooden floor in her kitchen. The floor is a rectangle measuring 4 m by 4.6 m. The wooden floorboards Fiona has chosen are 1 m long and 0.2 m wide.
How many floorboards does Fiona need?

3
A02
A03
Trevor is going to paint some doors in his house. Each door is a 2 m by 0.9 m rectangle and he is going to paint both sides of each door.
Each tin of paint that he is going to use covers 8 m². Trevor wants to paint 20 doors.
How many tins of paint does he need?

4
A02
A03
A tortoise has an enclosure in a garden.
It is in the shape of a 4.5 m by 2 m rectangle.
The tortoise's owner builds a new square enclosure with the same area.
What is the side length of the new enclosure?

5
A02
A03
Naomi and Gary are redoing their bathroom. It is to have new tiles on the wall and laminate flooring.
The widths of the tiled walls are 2.5 m and 3.5 m.
The height of both tiled walls is 2.8 m. The floor is a rectangle measuring 2.5 m by 3.5 m, but the bath takes up a rectangle measuring 1.7 m by 0.75 m.

a How many square tiles, each with side length 30 cm, must be bought for the walls?

b What area of flooring is needed?

c Laminate flooring costs £7.98 per square metre. How much will it cost to cover the bathroom floor?

C 6
A02

The diagram shows a piece of blue paper with a corner removed.

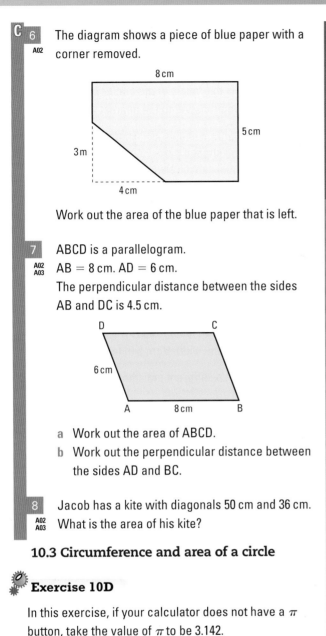

8 cm

5 cm

3 m

4 cm

Work out the area of the blue paper that is left.

7
A02
A03

ABCD is a parallelogram.

AB = 8 cm. AD = 6 cm.

The perpendicular distance between the sides AB and DC is 4.5 cm.

D   C

6 cm

A   8 cm   B

a Work out the area of ABCD.

b Work out the perpendicular distance between the sides AD and BC.

8
A02
A03

Jacob has a kite with diagonals 50 cm and 36 cm. What is the area of his kite?

10.3 Circumference and area of a circle

Exercise 10D

In this exercise, if your calculator does not have a π button, take the value of π to be 3.142.

Give answers correct to 3 significant figures unless a question says differently.

D 1

Work out the circumference of a circle with diameter:

a 6 cm

b 15.9 mm

c 6.5 cm

d 30 cm

e 12.8 m

ResultsPlus

Watch Out!

Be careful when dividing by 2π on a calculator. It is best to use brackets.

D 2

The radius of the largest disc on a machine is 26 cm.

a Work out the circumference of the disc.

The smallest disc on the machine has a radius of 18 cm.

b Work out how much longer the circumference of the largest disc is than the circumference of the smallest disc.

C 3

The Olympic Stadium for the 2012 Olympics is planned to have a circumference of 900 metres. Work out the radius of a circle with this circumference.

4

The diagram shows a rug in the shape of a semicircle of radius 75 cm.

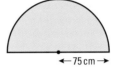

←75 cm→

Work out the perimeter of the rug.

5
A02

The minute hand of a clock is 11 cm long.

The clock chimes every $\frac{1}{4}$ hour.

a Work out the distance moved by the point end of the hand between chimes.

b Work out the distance moved by the point end of the hand when the clock has chimed three times.

B 6

The circular surface of a cake stand has a radius of 25 cm.

a Work out the circumference of the cake stand.

The circumference of a cake is 126 cm.

The cake is placed on the cake stand, exactly in the centre.

b Work out the distance from the edge of the cake to the edge of the cake stand.

7
A02
A03

The wheel of a wheelbarrow turns 80 times when it is pushed a distance of 70 m.

Work out the radius of the wheel.

Give your answer in cm correct to the nearest cm.

Exercise 10E

In this exercise, if your calculator does not have a π button, take the value of π to be 3.142.

Give answers correct to 3 significant figures unless the question says differently.

D 1 Work out the area of a circle with radius:

    a 6 cm   b 17.4 cm   c 25.6 mm

    d 9.27 cm   e 16.2 m

2 Work out the area of a circle with diameter:

    a 26 cm   b 3.8 cm   c 0.85 m

    d 55.8 mm   e 16.39 cm

C 3 A commemorative coin is made of two different
A02
A03 metals. One metal is contained within the other, as shown in the diagram.

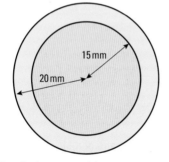

    a What is the area of the central metal?

    b What is the area of the outer metal?

4 The diagram represents the name badges
A02
A03 used by a supermarket. Each badge is a rectangle with semicircular ends. The rectangle measures 5 cm by 3.6 cm. The semicircles have a diameter of 3.6 cm.

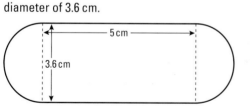

    a Work out the area of one badge.

The plastic from which the badges are cut costs 3p per cm².

    b Work out the cost of the plastic used to make one badge.

5 A circular floor has radius 2 m. The floor is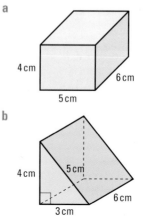
A02
A03 to be painted with special floor paint. The floor is to be painted twice. Each tin of paint covers 6 m². Work out how many tins of paint will need to be bought.

B 6 A rectangular card measures 30 cm by 20 cm.
A02
A03 Two identical circles are cut out of the card. The area left is 443 cm².
What is the radius of the circles?
Give your answer to the nearest whole number.

7 The diagram shows a
A02
A03 tile. The shaded parts are equal semicircles. What is the area of white on the tile?

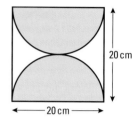

10.4 Drawing 3D shapes

Exercise 10F

1 Use isometric paper to draw a cuboid with height 3 cm, width 6 cm and length 4 cm.

2 Sketch six different nets that will make the same cuboid.

3 Here are the nets of some 3D shapes. Identify the shapes.

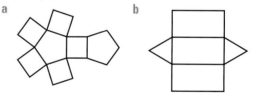

4 Draw an accurate net for each of these.

    a

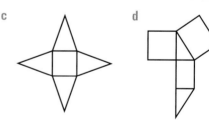

    b

10.5 Elevations and plans

Exercise 10G

D 1 Draw the elevations and plans of these shapes.

a

front

b

c

3 m
4 m
6 m

d

6 cm
7 cm
3 cm

e

2 cm
4 cm
6 cm
2 cm
4 cm
6 cm

f

3 cm
7 cm

g

6 cm
5 cm

D 2 Sketch the shapes represented by these elevations and plans.

a

plan

front elevation   side elevation

b

plan

front elevation   side elevation

c

plan

front elevation   side elevation

10.6 Volume of a cuboid

Exercise 10H

D 1 These shapes are made from cuboids. Work out the volumes of the shapes.

a

6 cm
140 mm
4 cm
12 cm
8 cm

b

5 cm
3 cm
9 cm
11 cm
11 cm

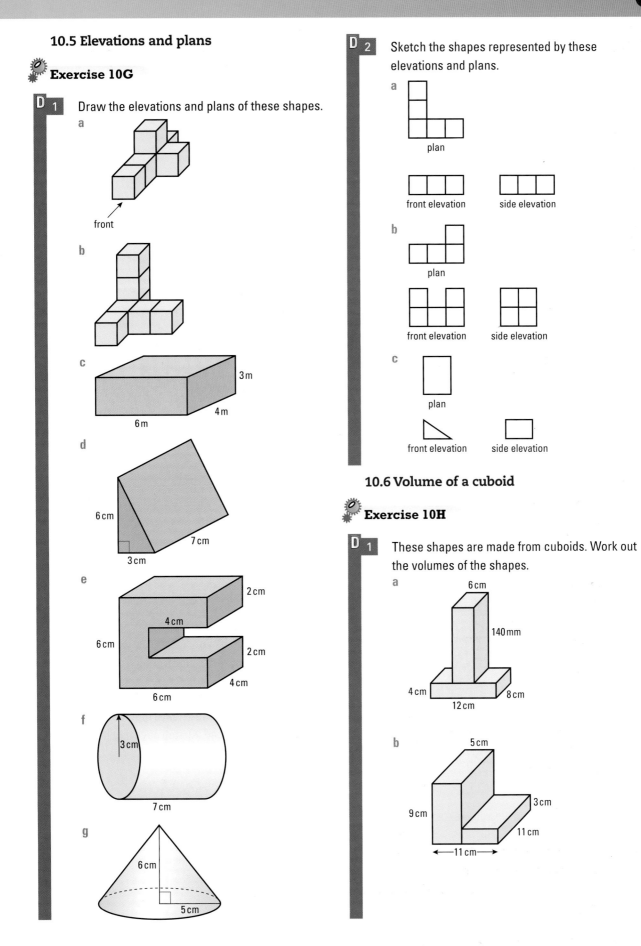

D

c

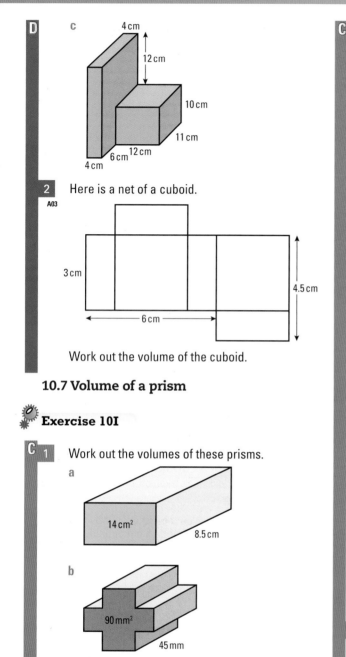

2 Here is a net of a cuboid.

A03

Work out the volume of the cuboid.

10.7 Volume of a prism

Exercise 10I

C 1 Work out the volumes of these prisms.

a

14 cm²

8.5 cm

b

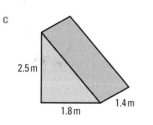

90 mm²

45 mm

c

2.5 m

1.8 m   1.4 m

d

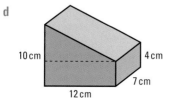

10 cm   4 cm

7 cm

12 cm

C 2 Work out the volumes of these prisms.

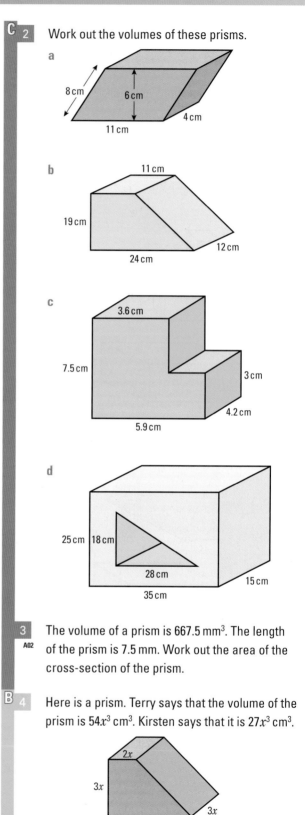

a

8 cm   6 cm   4 cm

11 cm

b

11 cm

19 cm

24 cm   12 cm

c

3.6 cm

7.5 cm

3 cm

4.2 cm

5.9 cm

d

25 cm   18 cm

28 cm

35 cm   15 cm

3 The volume of a prism is 667.5 mm³. The length
A02 of the prism is 7.5 mm. Work out the area of the
cross-section of the prism.

B 4 Here is a prism. Terry says that the volume of the
prism is $54x^3$ cm³. Kirsten says that it is $27x^3$ cm³.

2x

3x

3x

4x

Who is correct? You must show your working.

B 5 The diagram shows a triangular prism.
The volume of the prism is $45x^3$ cm³.

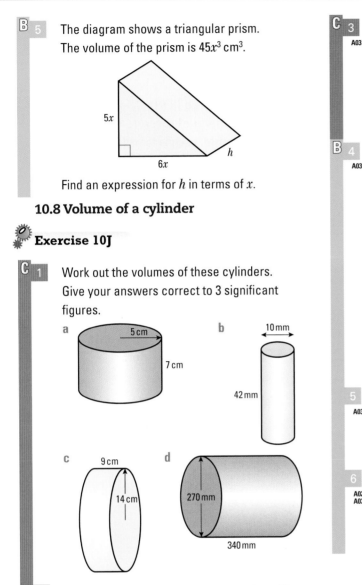

$5x$

$6x$

h

Find an expression for h in terms of x.

10.8 Volume of a cylinder

Exercise 10J

C 1 Work out the volumes of these cylinders.
Give your answers correct to 3 significant figures.

a 5 cm 7 cm

b 10 mm 42 mm

c 9 cm 14 cm

d 270 mm 340 mm

2 Work out the volumes of these cylinders. Give your answers in terms of π.

a 7 cm 12 cm

b 24 cm 7.5 cm

c 0.55 m 0.8 m

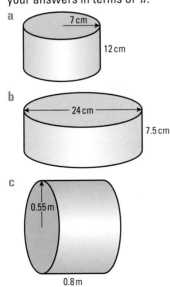

C 3 This child's brick has a semicircular cross-section of diameter 2.6 cm.
The length of the brick is 6 cm.
Work out the volume of the brick. Give your answer in terms of π.

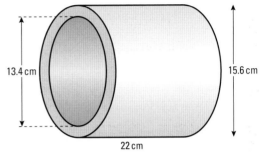

6 cm

2.6 cm

B 4 This piece of pipe has external diameter 15.6 cm, internal diameter 13.4 cm and length 22 cm.

13.4 cm   15.6 cm

22 cm

Work out the volume of the pipe.
Give your answer correct to 1 decimal place.

5 A cylindrical can has a radius of 3.5 cm and a volume of 416 cm³.
Work out the height of the can.
Give your answer correct to 1 decimal place.

6 A beaker is in the shape of a cylinder with radius 6.6 cm. It is filled to a height of 14.8 cm.
Work out the volume of liquid in the beaker.
Give your answer correct to 3 significant figures.

11 Averages and range

Key Points

- **average:** a single value that is representative of all the values in a set of quantitative data.
- **mode:** the most frequent value in a set of data.
- **median:** the middle value, or halfway between two middle values, in an ascending set of data.
- **mean:** the most common average used.

$$\text{mean} = \frac{\text{sum of values}}{\text{number of values}} = \frac{\Sigma x}{n},$$

for a sample of n values of x.

- **frequency table:** a table of a set of observations showing how frequently each event or quantity occurs.
- **modal class of grouped data:** the class interval with the highest frequency.
- **median of grouped data:** the class interval in which the median falls.
- **range:** describes the spread of the data, range = the highest value – the lowest value.
- **quartiles:** split the data into four parts

Lowest value Q_1 Q_2 Q_3 Highest value

25% of data   25% of data   25% of data   25% of data

- **lower quartile (Q_1):** quarter of the way through the data
- **second quartile (Q_2):** halfway though the data (the median)
- **upper quartile (Q_3):** three-quarters of the way through the data

- **interquartile range (IQR):** $Q_3 - Q_1$
- **finding the mean of discrete data in a frequency table:** use the formula

$$\text{mean} = \frac{\Sigma f \times x}{\Sigma f}$$

where f is the frequency, x is the variable and Σ means 'the sum of'.

- **estimating the mean of grouped data in a frequency table:** use the formula

$$\text{mean} = \frac{\Sigma f \times x}{\Sigma f}$$

where f is the frequency, x is the class midpoint and Σ means 'the sum of'.

- **finding quartiles for a set of n data values arranged in ascending order:**
  - Q_1 is the $\left(\frac{n+1}{4}\right)$th value
  - Q_2 is the $\left(\frac{n+1}{2}\right)$th value
  - Q_3 is the $\left(\frac{3(n+1)}{4}\right)$th value

Measure	Advantages	Disadvantages
MODE Use the mode when the data are non-numeric or when asked to choose the most popular item.	Extreme values (outliers) do not affect the mode. Can be used with qualitative data.	There may be more than one mode. There may not be a mode, particularly if the data set is small.
MEDIAN Use the median to describe the middle of a set of data that does have an extreme value.	Not influenced by extreme values.	Not as popular as mean. Actual value may not exist.
MEAN Use the mean to describe the middle of a set of data that does not have an extreme value.	Is the most popular measure. Can be used for further calculations. Uses all the data.	Affected by extreme values.

11.1 Finding the mode and median

Exercise 11A

Questions in this chapter are targeted at the grades indicated.

1   During the cricket season a local team played 27 matches. A batsman scored the following numbers of runs in each match.

15	16	27	24	38	38	40	16
33	20	11	35	33	10	42	15
24	20	24	25	21	40	24	24
14	30	41					

Find the mode.

2   A delivery woman buys fuel every day at her local garage. The following are the quantities she buys in one particular week.

Monday 22 litres   Tuesday 20 litres
Wednesday 15 litres   Thursday 14 litres
Friday 18 litres   Saturday 18 litres
Sunday 22 litres

Find the median amount she buys.

3   Here are the ages of the members of a snooker club.

15	16	14	16	17	15	14	16
17	14	15	16	14	15	16	17
16	14	16	16	15	15	14	14

a   Find the mode.   b   Find the median age.

4   A coach travels from Birmingham to Penzance. It makes seven stops along the way. The number of passengers on each stage of the journey is shown below.

Birmingham to Worcester   40
Worcester to Bristol   45
Bristol to Plymouth   50
Plymouth to Bodmin   52
Bodmin to Wadebridge   47
Wadebridge to Newquay   45
Newquay to St Ives   38
St Ives to Penzance   24

The coach company wants to find the median number of passengers on the coach.

11.2 Calculating the mean

Exercise 11B

1   An online grocery service has 12 deliveries to make in the Wakefield area in one morning. The number of boxes of groceries delivered at each house is as follows:

4 3 5 2 2 2 3 3 4 3 2 3

Work out the mean number of boxes delivered per house.

2   A high school has the following numbers of boys and girls in each year.

Year	Boys	Girls
7	129	126
8	135	130
9	156	150
10	152	154
11	152	145

a   Find the mean number of boys per year.
b   Find the mean number of girls per year.

3   Sally gets some casual work at an asparagus farm for five days. She picks an average of 315 bunches of asparagus per day.
Work out how many bunches of asparagus Sally picks in total over the five days.

4   In a Test match, the following numbers of runs were scored by one team.

11   23   20   46   62   74   35   52   45   11   6

a   Find the mode.
b   Find the median number of runs.
c   Find the mean number of runs.

11.3 Using the three types of average

Exercise 11C

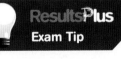

1   Six friends buy themselves shoes in the January sales. They spend the following amounts of money.

Exam Tip
When comparing averages, look at how well each average represents the numbers as a whole, and give reasons why they would or would not be representative.

£26   £38   £45   £99   £38   £54

a   Work out the mean, mode and median values.
A03 b   Which average would be the best to use to describe the amount of money they spent?

D 2 Write down one advantage and one disadvantage of using the median as an average.

3 A café sells the following numbers of lunches during one week.

30   38   30   40   65   122   134

a Write down the modal number of lunches sold.

b Work out the median number of lunches sold.

c Work out the mean number of lunches sold.

A03 d The owner puts an advertisement in the local paper. He wants to use an average to encourage customers to eat in his café. Which should he use?

Give a reason for your answer.

11.4 Using frequency tables to find averages

Exercise 11D

D 1 The table shows information about the number of children in each house on a street.

Number of children (x)	Frequency (f)	Frequency × number of children (f × x)
0	5	
1	7	
2	10	
3	5	
4	2	
5	0	
6	1	
Total		

a Copy and complete the table.

b Write down the mode of these data.

c Work out the median number of children.

d Work out the mean number of children.

2 The table shows information about the numbers of elastic bands in each packet of a box containing 50 packets.

Number of elastic bands (x)	51	52	53	54	55	56	57
Frequency (f)	5	6	9	15	8	5	2

a Write down the mode of these data.

b Work out the median number of elastic bands.

c Work out the mean number of elastic bands.

D 3 A chain store runs a management training scheme for university graduates. The age range of recruits to the scheme is shown in the table.

Ages of recruits (x years)	21	22	23	24	25	26
Frequency (f)	9	10	10	3	3	3

a Write down the mode of these data.

b Work out the median age.

c Work out the mean age.

11.5 Modal class and median of grouped data

Exercise 11E

D 1 A dentist uses a bur in his drill. The bur consists of a shank and a drill bit. The table shows the sizes of different burs kept in the dentist's surgery.

Length (l mm)	Frequency (f)
$20 \leqslant l < 21$	15
$21 \leqslant l < 22$	22
$22 \leqslant l < 23$	38
$23 \leqslant l < 24$	42
$24 \leqslant l < 25$	20
$25 \leqslant l < 26$	10
$26 \leqslant l < 27$	3

a Write down the modal class.

b Find the class into which the median falls.

C 2 Richard runs a retail business through an auction website. His weekly income for the past year is shown in the following grouped frequency table.

Weekly income (£s)	Frequency (f)
£201–£240	5
£241–£280	19
£281–£320	18
£321–£360	10

a Write down the modal class.

b Find the class into which the median falls.

C **3** The table shows the weight of gold wasted when 50 crowns are made by a dental technician.

Weight (w g)	Frequency (f)
$0.05 \leqslant w < 0.10$	12
$0.10 \leqslant w < 0.15$	14
$0.15 \leqslant w < 0.20$	16
$0.20 \leqslant w < 0.25$	8

a Write down the modal class.
b Find the class into which the median falls.

11.6 Estimating the mean of grouped data

Exercise 11F

C **1** The following grouped frequency table shows the ages of members of a salsa club.

Age range (years)	Frequency
18–26	5
27–35	11
36–44	14
45–53	10
54–62	9
63–71	1

Work out an estimate for the mean age of the members.

2 The practice manager of a health centre recorded the length of time, in minutes, that patients had to wait to see a doctor. The results are shown in the grouped frequency table below.

Waiting time (t minutes)	Frequency
$0 \leqslant t < 5$	12
$5 \leqslant t < 10$	38
$10 \leqslant t < 15$	23
$15 \leqslant t < 20$	7

Work out an estimate for the mean waiting time of the patients to the nearest minute.

C **3** A curry house kept a record of the distances of home deliveries it made in one week. The results are shown in the grouped frequency table.

Distance (d km)	Frequency
$0 \leqslant d < 3$	16
$3 \leqslant d < 6$	22
$6 \leqslant d < 9$	17
$9 \leqslant d < 12$	12
$12 \leqslant d < 15$	3

Work out an estimate for the mean distance of the deliveries. Give your answer correct to 3 significant figures.

11.7 Range, quartiles and interquartile range

Exercise 11G

B **1** Write down another name for Q_3.

2 Eleven teenagers were asked to record the amount of time they spent on their mobile phones one evening. Their times, in minutes, were:

14 21 30 42 48 52 55 60 63 72 77

a Write down Q_1, Q_2 and Q_3 for these data.
b Work out the interquartile range.
c Work out the range.

3 A research assistant counted the number of butterflies on a buddleia bush at midday over a period of 15 days. His findings are listed below.

11 20 6 15 12 15 8 16 19 8
9 10 18 23 7

a Write down Q_1, Q_2 and Q_3 for these data.
b Work out the interquartile range.
c Work out the range.

4 The number of magazines sold per day in a newsagent was recorded over 17 days. The results are shown below.

23 43 45 39 58 34 56 43 52
43 30 34 70 18 46 60 27

a Write down Q_1, Q_2 and Q_3 for these data.
b Work out the interquartile range.
c Work out the range.

12 Constructions and loci

12.1 Constructing triangles

Exercise 12A

Questions in this chapter are targeted at the grades indicated.

D **1** Here is a sketch of triangle ABC.

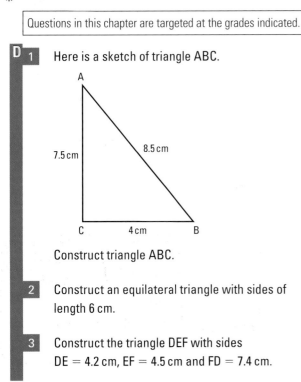

Construct triangle ABC.

2 Construct an equilateral triangle with sides of length 6 cm.

3 Construct the triangle DEF with sides DE = 4.2 cm, EF = 4.5 cm and FD = 7.4 cm.

D **4** Here is a sketch of the quadrilateral WXYZ.

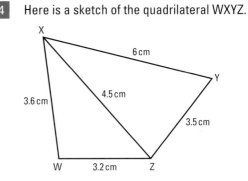

Make an accurate drawing of quadrilateral WXYZ.

5 The rhombus STUV has sides of length 5 cm.
The diagonal SU = 8 cm.
Make an accurate drawing of the rhombus STUV.

6 Explain why it is not possible to construct a triangle with sides of length 5 cm, 11 cm and 5 cm.

12.2 Perpendicular lines

Exercise 12B

1. Draw line segments of length 12 cm and 6 cm. Using a straight edge and a pair of compasses, construct the perpendicular bisector of each of these line segments.

2. Draw these lines accurately, and then construct the perpendicular from the point P.

a

b

3. Draw a vertical line segment CD, a point to the left of it, E, and a point to the right of it, F. Construct the perpendicular from E to CD, and from F to CD.

12.3 Constructing and bisecting angles

Exercise 12C

1. Copy the diagrams and construct the bisector of the angle XYZ.

a

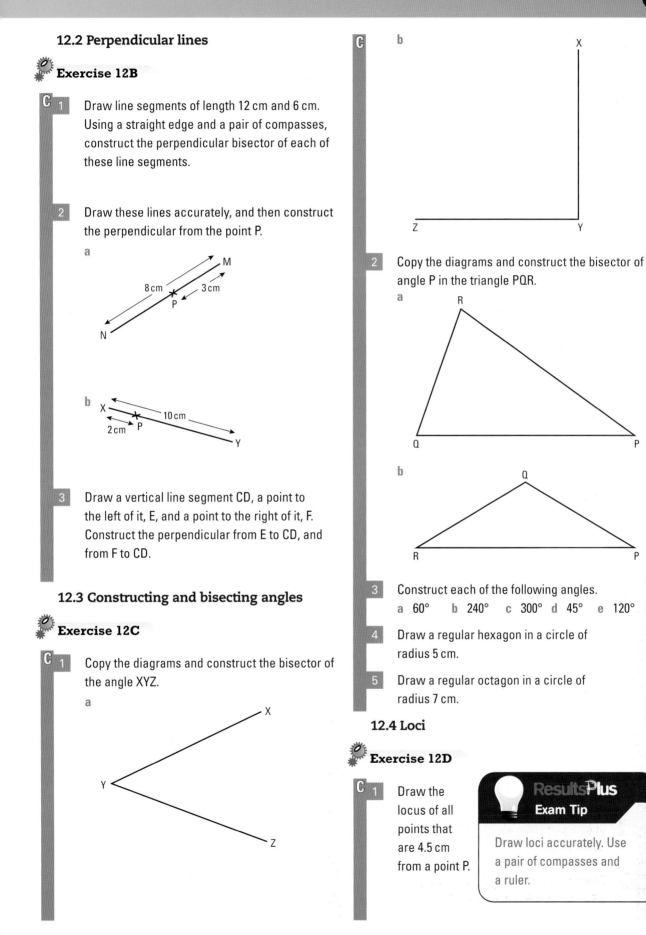

b

2. Copy the diagrams and construct the bisector of angle P in the triangle PQR.

a

b

3. Construct each of the following angles.
   a 60° b 240° c 300° d 45° e 120°

4. Draw a regular hexagon in a circle of radius 5 cm.

5. Draw a regular octagon in a circle of radius 7 cm.

12.4 Loci

Exercise 12D

1. Draw the locus of all points that are 4.5 cm from a point P.

ResultsPlus

Exam Tip

Draw loci accurately. Use a pair of compasses and a ruler.

C **2** Mark two points X and Y approximately 8 cm apart. Draw the locus of all points that are equidistant from X and Y.

3 Draw two lines AB and BC, so that the angle ABC is obtuse. Draw the locus of all points that are equidistant between the two lines AB and BC.

4 Draw the locus of a point that moves so that it is always 2 cm from a line 6.5 cm long.

12.5 Regions

Exercise 12E

C **1** Shade the region of points which are less than 3 cm from a point P.

2 Mark two points, A and B, roughly 4.5 cm apart. Shade the region of points which are closer to B than to A.

3 Shade the region of points which are less than 3.2 cm from a line 5.5 cm long.

4 Draw two lines XY and YZ, so that the angle XYZ is obtuse. Shade the region of points which are closer to XY than to YZ.

5 An equestrian centre has a horse-walking machine. The machine consists of a 4.5 m arm, one end of which is fixed at the centre of the room. A horse is attached to the other end of the arm so that, as the machine rotates, the horse is walked around the room. As well as walking forwards, the horse has the freedom to move up to 1 m to its left or 1 m to its right.
Show the region of points a horse on the machine can walk in.

12.6 Scale drawings and maps

Exercise 12F

D **1**
A02
A03
This is a plan of a new adventure park.
Key: C = climbing equipment
         S = swimming pool
         T = tree-top walk
         R = refreshment area

D The real distance between the swimming pool and the climbing equipment is 600 m.

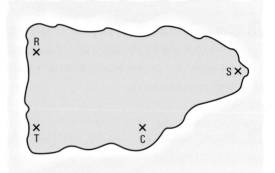

a Find the scale of the map in the form 1 cm represents n m.
b What is the real distance of the refreshment area from:
   i C   **ii** T   **iii** S?

2 On a map of England, 1 cm represents 25 km.
a The distance between Brighton and Oxford is 170 km. Work out the distance between Brighton and Oxford on the map.
b On the map, the distance between London and Manchester is 12.8 cm. Work out the real distance between London and Manchester.

3
A02
D is a distribution centre that serves two supermarkets, S and T. This map shows the distances between D, S and T. It is *not* accurately drawn.

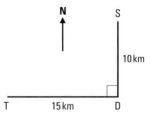

a Using a scale of 1 : 500 000, draw this map accurately.
b What is the real distance between the two supermarkets?
c Measure the bearing of S from T.

D 4 This is a sketch of Adam's dining room. It is *not* drawn to scale.

Draw an accurate scale drawing on cm squared paper of Adam's dining room. Use a scale of 1 : 100.

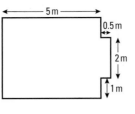

C 5 A new shopping centre has a length of 42 m. A model of the shopping centre has a length of 60 cm.

a Find, in the form 1 : n, the scale of the model.

b The height of the shopping centre is 10.5 m. Work out the height of the model.

C 6 The distance between Reading and Birmingham is 168 km. On a map, the distance between Reading and Birmingham is 8.4 cm.

a Write down the scale of the map as a ratio.

b The distance between Birmingham and London is 188 km. How far is that on the map in cm?

13 Linear equations

Key Points

- **an equation:** has an equals sign and a symbol or letter that represents an unknown number, e.g. $2p - 10 = 9$.
- **an expression:** a collection of terms and variables, e.g. $2p - 10$.
- **an identity:** an equation that is true for all values of the variables, e.g. $2p - 10 = 2(p - 5)$.
- **a formula:** where one variable is equal to an expression in a different variable(s), e.g. $C = 2p - 10$.

- **solving simple linear equations:** apply operations to both sides of the equation until the unknown appears on its own on one side of the equation only.
  - solutions can be whole numbers, fractions, decimals, or negative numbers
  - follow the rules of BIDMAS from Chapter 1
  - collect like terms
  - eliminate fractions by multiplying each term by the LCM of the denominators

13.1 Solving simple equations

Exercise 13A

Questions in this chapter are targeted at the grades indicated.

Solve

1 $2x + 9 = 15$

2 $3y - 1 = 17$

3 $2a + 11 = 14$

4 $5g - 4 = 7$

ResultsPlus
Watch Out!

Always apply the same operation to **both** sides of the equation.

5 $15 + 4k = 7$ **6** $7 - 2m = 12$

7 $15 = 3n - 6$ **8** $5 = 3 + 8b$

9 $4 - 10r = 34$ **10** $21 = 1 - 5p$

13.2 Solving linear equations containing brackets

Exercise 13B

Solve

D 1 $8(a + 3) = 24$ **2** $8(b + 4) = 40$

3 $3(x - 2) = 5$ **4** $4(2h - 1) = 16$

5 $15 = 4(t + 5)$ **6** $5(1 - y) = 10$

7 $4(7 - 4s) = 20$ **8** $2 = 5(1 + 3k)$

9 $5(2c + 6) + 1 = 26$ **10** $13 = 7 - 4(6 - 3d)$

13.3 Solving linear equations with the unknown on both sides

Exercise 13C

Solve

D 1 $4x + 4 = 2x + 9$

2 $2y - 12 = 10y + 4$

3 $6z - 11 = 4z - 3$

4 $d + 6 = 5d + 22$

ResultsPlus
Exam Tip

Collect the terms in x on the side of the equation that gives them a positive coefficient.

5 $4 - 2u = 3 - 3u$ **6** $2 - 8v = 2v + 9$

7 $x + 13 = 3(x - 8)$

C 8 $6 + 2q = 3(q + 2) + 2$

9 $3(4 - 5x) + 2(5x + 2) = 3 + x$

10 $j - 2(1 - 2j) = 3(j - 3) + 5$

13.4 Solving linear equations containing fractions

Exercise 13D

Solve

ResultsPlus
Exam Tip

Always try to remove the fraction first.

B 1 $\dfrac{x}{4} + 5 = 6$

2 $\dfrac{h + 2}{5} = 7$

3 $\dfrac{p}{3} + \dfrac{p}{4} = 14$ **4** $\dfrac{c}{4} + 1 = \dfrac{x - 4}{6}$

5 $3(3x - 5) = \dfrac{4x + 1}{3}$

6 $2\left(\dfrac{m}{4} - 4\right) = 16$ **7** $\dfrac{1}{4k} + \dfrac{1}{3k} = -7$

A

8 $\dfrac{10 - 4x}{6} + \dfrac{12 + 6x}{2} = 12$

9 $\dfrac{2 - 7x}{3} = 5 - \dfrac{x + 4}{9}$

10 $\dfrac{5n + 1}{2} - \dfrac{3n - 10}{3} = \dfrac{6 - 2n}{4}$

13.5 Setting up and solving simple linear equations

Exercise 13E

D

1 Steve thinks of a number. He multiplies the number by 4 and then subtracts 3. His answer is 17. If x is the number that Steve was thinking of, work out the value of x.

2 A02 Naomi's brother, Dave, is 3 years younger than she is. The sum of their ages is 81. How old is Dave?

C 3 A02 In an election, candidate A scored m votes. Candidate B scored $\frac{3}{10}$ of A's votes. Candidate C scored $\frac{2}{5}$ fewer votes than A. The total number of votes cast in the election was 3800. How many votes did each candidate get?

4 A03 A rectangle measures $(3x + 1)$cm by $\frac{1}{4}(x + 2)$cm.

$3x + 1$

$\frac{1}{4}(x + 2)$

If the perimeter of the rectangle is 60 cm, what is the value of x?

5 A03 Miriam works flexitime. She must work 110 hours in a 5-week period. She works t hours for each of 4 weeks. In the 5th week, she works an extra $3\frac{1}{2}$ hours. Work out the number of hours Miriam works in the 5th week.

B 6 The diagram shows an isosceles trapezium ABCD. The lengths of the sides are given in centimetres and AB = CD.

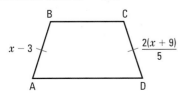

a Write down an equation in terms of x.
b Work out the lengths of AB and CD.

B 7 A03 Phil checks his water bill. He has used g cubic metres of water.

Cost of the first 52 units at 70.72p per cubic metre
= £ _____

Cost of remaining units at 68.8p per cubic metre
= £ _____

Total cost of cubic metres used = £85.47

Work out the number of cubic metres of water Phil used. Give your answer correct to one decimal place.

13.6 Distinguishing between 'equation', 'formula', 'identity' and 'expression'

Exercise 13F

Write down whether each of the following is an expression, an equation, an identity or a formula.

> **ResultsPlus**
> **Exam Tip**
>
> An expression is the only one without an '=' sign.

1 $h + 2h + h = 4h$

2 $6x - \dfrac{y}{2}$

3 $d = st$

4 $a = \dfrac{v_2 - v_1}{t}$

5 $2z^2 + 3z$

6 $4p + 2 = 10$

7 $4gh + 5gh = 9gh$

8 $x^2 + 7 = 32$

9 $V = \dfrac{P}{I}$

D 10 $7 + m(m + 3) - 2 = m^2 + 3m + 5$

11 $3a + 2 = 5(a - 2)$

C 12 $\dfrac{d^{10}}{d^4} = d^6$

14 Percentages

Key Points

- **percentage (%):** a quantity out of 100. Can also be written as a decimal or a fraction.
- **compound interest:** interest paid on the total amount including previous interest.
    - amount after n years = original amount $\times$ multiplier^n
- **finding percentages of quantities:**
    - write the percentage as a fraction or decimal, and then multiply the fraction or decimal by the quantity
    - find 10% or 1% of the amount then build up the percentage
- **increasing an amount by a percentage:** either work out the percentage of the amount and add it to the original number, or use a multiplier.

- **decreasing an amount by a percentage:** either work out the percentage of the amount and subtract it from the original number, or use a multiplier.
- **writing one quantity as a percentage of another quantity:** write the first quantity as a fraction of the second quantity, then convert the fraction to a percentage.
- **finding the percentage profit or loss:** use the formula, percentage profit (or loss)
$$= \frac{\text{profit (or loss)}}{\text{original amount}} \times 100\%$$
- **calculating reverse percentages:** if you know the final amount after a percentage increase or decrease you can calculate the original amount by dividing by the multiplier.

14.1 Working out a percentage of a quantity

Exercise 14A

Questions in this chapter are targeted at the grades indicated.

1 Work out

a 20% of £700

b 45% of 60 kg

c 15% of £42.60

d 5% of 82 kg

e 30% of £30

f 40% of £350

g 5.2% of 120 km

h $17\frac{1}{2}$% of £400

ResultsPlus
Watch Out!

When the answer is an amount of money, remember to write it to two decimal places.

2 There are 250 flamingos in a flock. 12% of the flamingos are chicks.
How many of the flamingos are chicks?

3 Katie invests £450. The interest rate is 3% per year.
How much interest will she receive after 1 year?

4 A bookshop has 5200 books in stock. 28% of the books are romance novels.
How many of the books in the shop are romance novels?

D **5**
A02 A03
There are 146 students in Year 10. Of these students, 76 are boys. 50% of the boys and 60% of the girls are going on a theatre trip.
What fraction of the Year 10 students are going on the theatre trip?
Give the fraction in its simplest form.

14.2 Finding the new amount after a percentage increase or decrease

Exercise 14B

1 Write down the single number you can multiply by to work out an increase of

a 72%   b 4%   c 12%

d 20%   e 14.3%   f $17\frac{1}{2}$%

g 61%   h 1.63%

2 a In order to increase an amount by 30%, what single number should you multiply by?

b A rail company puts up its ticket prices by 30%. What is the new price of a ticket that previously cost £20.60?

D 3 The table shows the prices of three items on an auction website 24 hours before the auctions ended. By the end of the auctions, the price of each item had increased by 3.2%.

Mobile phone	£140
Hi-fi	£320
1st edition book	£330

Work out the selling price of each item.

4 Kwame puts £800 into a bank account. At the end of one year 2.5% interest is added. How much is in his account at the end of 1 year?

5
a Increase £160 by 40%.
b Increase 84 kg by 75%.
c Increase 3.2 m by 18%.
d Increase £1480 by 12.5%.
e Increase 236 cm by 4%.

Exercise 14C

1 Write down the single number you can multiply by to work out a decrease of
a 9%   b 60%   c 23%
d 14%   e 6.5%   f $4\frac{1}{2}$%
g $5\frac{3}{4}$%   h 0.6%

D 2 In a designer clothes sale all prices are reduced by 12%.
Work out the sale price of each of the following:
a a dress that normally costs £500
b a leather jacket that normally costs £1100
c a scarf that normally costs £40.

3 James has built up an ability level of 86 on his games console. His little brother then plays using James' avatar and his ability level goes down by 5%. What is James' ability level now?

C 4 A plasma television normally costs £950. It is reduced by 17%. How much will the television now cost?

5 Sam buys a motorbike for £8500. The value of the motorbike depreciates by 15% each year. Work out the value of the bike at the end of:
a 1 year
b 2 years.

14.3 Working out a percentage increase or decrease

Exercise 14D

D 1 Write:
a £4 as a percentage of £16
b 2p as a percentage of 10p
c 4 kg as a percentage of 14 kg
d 9 mm as a percentage of 90 mm
e 90p as a percentage of £5.40
f 15 cm as a percentage of 9 m
g 48 minutes as a percentage of 1 hour
h 12 mm as a percentage of 6 cm.

2 Martina scored 62 out of 80 in a Science test. Work out her score as a percentage.

3 Tao picked 120 apples. 48 of them were red. What percentage of the apples were red?

4 A 50 g flapjack contains 8 g of protein, 28 g of carbohydrates, 9 g of fat and 5 g of fibre. What percentage of the serving is:
a protein   b carbohydrates
c fat   d fibre?

Exercise 14E

C 1 Calculate the percentage increase or decrease to the nearest 1%:
a £32 to £48
b 12.5 cm to 15 cm
c 55 kg to 36.5 kg
d 3 minutes to 165 seconds.

2 A shop reduces, the price of a kettle from £36 to £28.35. Work out the percentage reduction.

3 Mandie bought 60 cherry pies for £45. She sold all the pies for £1.25 each at fundraising event. Work out her percentage profit.

C **4** * Gary owns three garden centres. When he looks
A03   at the number of items sold in the first half of the
       year and the number of items sold in the second
       half of the year, he finds that centre F has the
       greatest number of sales.

Garden centre	Items sold Jan–Jun	Items sold Jul–Dec
F	10 792	11 248
G	9765	10 320
H	8914	9688

Gary says that this must mean centre F has the
greatest percentage increase. Is he correct?
Explain your answer.

5 Lucy buys three crates, each containing 🅕🅢
A02   48 cartons of fruit juice, at a discount
A03   supermarket. Each crate costs £12. She sells 110
      cartons at the school fair for 40p each.
      The rest of the cartons were sold at 30p each on
      a stall in the playground the next day.
      Work out the percentage profit or loss Lucy
      makes for the school.

6 A tennis club has 56 members each 🅕🅢
A03   paying £70 per year for membership. The club
      pays rent of £3850 per year to the local park for
      its courts. The council is increasing the rent for
      next year by 6.5%. Another member is joining the
      club for the next year.
      What is the minimum percentage increase that
      will have to be added to the membership fee
      to cover the increased rent? Give your answer
      correct to 1 decimal place.

14.4 Working out compound interest

Exercise 14F

D **1** Work out the multiplier as a single decimal
       number that represents:
       a an increase of 10% for 3 years
       b a decrease of 40% for 5 years
       c an increase of 5% followed by an increase
          of 3%
       d a decrease of 30% followed by a decrease
          of 25%.

C **2** £2000 is invested for 3 years at 4% 🅕🅢
       per annum compound interest. Work out the total
       amount in the account after 3 years.

B **3** Mr Greer buys a flat for £80 000. The value of
A02   the flat decreases in the first year by 5%. It then
A03   increases in value by 15% in the second year.
       a What decimal can you multiply the purchase
          price of the flat by to find its value after
          2 years?
       b What is the value of the flat after 2 years?

4 Adam says a decrease of 30% followed by a
A03   decrease of 20% is the same as a decrease of
      50%. Is Adam correct? Explain your answer.

5 Saskia puts £4000 in a savings account. 🅕🅢
      Compound interest is paid at a rate of 3% per
      annum. Saskia wants to leave the money in
      the account until there is at least £4500 in the
      account.
      Calculate the least number of years Saskia must
      leave her money in the savings account.

14.5 Calculating reverse percentages

Exercise 14G

B **1** A company awards its employees a pay 🅕🅢
A03   rise of 3%. The following week, Ellie earns
       £25 650.
       What did Ellie earn in the previous week?

2 The price of a laptop computer is £425. 🅕🅢
    This price includes Value Added Tax (VAT) at
    $17\frac{1}{2}$%. Work out the cost of the laptop before VAT
    was added.

3 A discount website is selling a home gym 🅕🅢
A03   for £510. The same home gym costs 32% more if
      bought from the manufacturer's website.
      What is the price on the manufacturer's site?

4 A job is advertised at a take-home wage of 🅕🅢
A03   £160.20 per week. What is the original pay if tax
      is deducted at a rate of 22%?

5 Over the past year, the number of trout in a lake
    has increased by 2.8%. There are currently 365
    trout in the lake.
    How many trout were there a year ago?

6 One year ago, Julian put some money he 🅕🅢
    inherited in a bank account. The account pays
    interest at a rate of 6% per annum. Julian now
    has £583 in his account.
    How much money did Julian inherit?

15 Graphs

- **line segment:** a line joining two points.
- **midpoint of a line:** halfway along the line.
- **gradient:** the slant of a line.
  - **gradient of a line** $= \dfrac{\text{change in } y\text{-direction}}{\text{change in } x\text{-direction}}$
  - **positive gradient:** a line that slants upwards from left to right
  - **negative gradient:** a line that slants downwards from left to right
  - steeper lines have larger gradients
- **y-intercept:** the value of y when $x = 0$. It is shown by the point $(0, c)$ where the graph crosses the y-axis.
- $y = mx + c$: the equation of a straight line where m is the gradient and c is the y-intercept.

- **distance–time graph:** a graph showing movement over time.
  - a slanting line shows movement
  - a horizontal line shows no movement
  - straight lines represent constant speed
  - the gradient gives the speed
- **drawing a straight line:** plot at least two points that fit the equation of the line.
- **finding the midpoint of a line:** the midpoint of the line segment AB between A(x_1, y_1) and B(x_2, y_2) is $\left(\dfrac{x_1 + x_2}{2}, \dfrac{y_1 + y_2}{2}\right)$.
- **finding the gradient of a line:** the gradient of the line through the points (x_1, y_1) and (x_2, y_2) is given by $m = \dfrac{y_2 - y_1}{x_2 - x_1}$.

15.1 Drawing straight-line graphs by plotting points

Exercise 15A

Questions in this chapter are targeted at the grades indicated.

1 Draw, on the same axes, the graphs of:

    a $x = -2$ b $y = 3$ c $x = 2.5$

    d $y = 0$ e $y = -2$ f $x = 0$

2 Write down the equation of each of these lines.

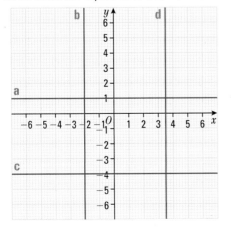

3 Write down the coordinates of the point where the following pairs of lines cross.

    a $x = 2, y = 1$ b $x = -2, y = 4$

    c $y = \frac{1}{2}, x = -1$

D **4**   **A02**   Work out the perimeter and area of a rectangle formed by the lines $x = 2$, $x = -2$, $y = 4$ and $y = -3$.

Exercise 15B

1   a   Copy and complete the table of values for $y = 5 - 2x$.

x	-2	-1	0	1	2	3	4
y			5		1	-1	

    b   Draw the graph of $y = 5 - 2x$. Use values of x from -2 to $+4$.

D **2**   a   Draw the graph of $y = 3x - 6$. Use values of x from -1 to $+4$.

    b   Write down the coordinates of the point where the graph intersects:

        i   the x-axis   ii   the y-axis.

    c   Use your graph to find:

        i   the value of y when $x = 1.5$

        ii   the value of x when $y = 6$.

3   **A03**   a   On the same axes, draw the graphs of $y = x$, $y = -x$, $y = -2x$ and $y = 3x$.

    b   What can you say about the graphs of all lines with equations of the form $y = mx + c$ with $c = 0$?

D 4   a   On the same axes, draw the graphs of
$y = 4x - 8$ and $y = 8 - 4x$.

      b   Write down the coordinates of the point
where the two graphs intersect.

C 5   A02 A03   Work out the area of the triangle formed by the
lines $y = 2x + 4$, $y = 0$ and $y = x$.

Exercise 15C

C 1   Draw the graph of $2x + 3y = 12$.

2   A03   a   On the same axes, draw the graphs of:

        i $x + y = 6$ ii $x + y = -3$ iii $x + y = 3$.

      b   What do you notice about the graphs you
have drawn?

3   A02   Find the coordinates of the point where the lines
$3x + 4y = 13$ and $x - y = 2$ intersect.

4   A02 A03   a   On the same axes, draw the graphs of
$2x + 5y = 8$ and $2y - 5x = 10$.

      b   What do you notice about the way the graphs
intersect?

      c   Find the equations of two other lines that
intersect in the same way.

15.2 Finding the midpoint of a line segment

Exercise 15D

D 1   Work out the coordinates of the midpoint of each
of the line segments shown on the grid.

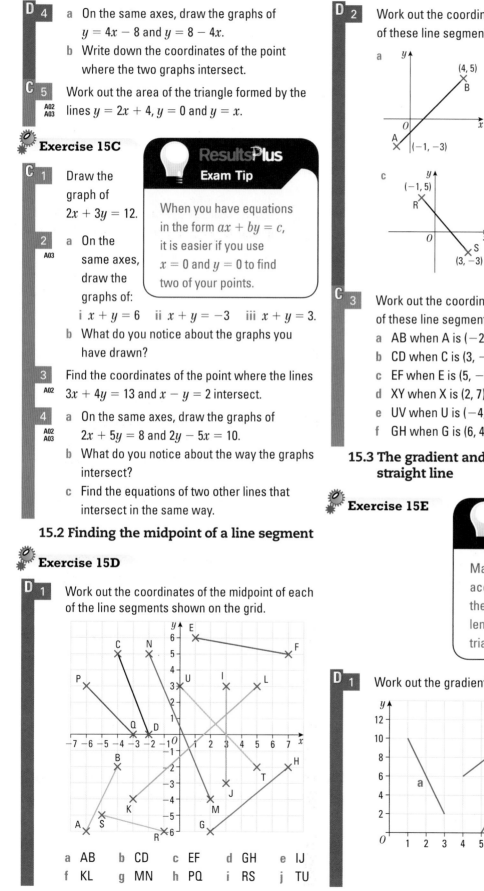

    a   AB   b   CD   c   EF   d   GH   e   IJ

    f   KL   g   MN   h   PQ   i   RS   j   TU

D 2   Work out the coordinates of the midpoint of each
of these line segments.

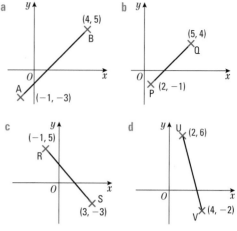

C 3   Work out the coordinates of the midpoint of each
of these line segments.

    a   AB when A is $(-2, -2)$ and B is $(9, 9)$

    b   CD when C is $(3, -5)$ and D is $(-6, 7)$

    c   EF when E is $(5, -8)$ and F is $(-3, 2)$

    d   XY when X is $(2, 7)$ and Y is $(-7, 1)$

    e   UV when U is $(-4, 3)$ and V is $(6, -7)$

    f   GH when G is $(6, 4)$ and H is $(-2, -6)$

15.3 The gradient and y-intercept of a straight line

Exercise 15E

D 1   Work out the gradient of each line.

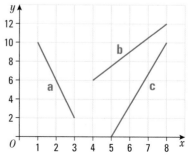

D

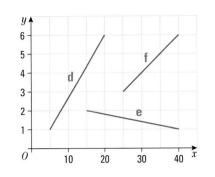

2 Work out the gradient and y-intercept of each straight line.

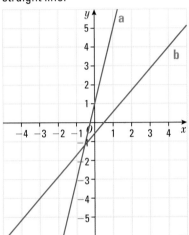

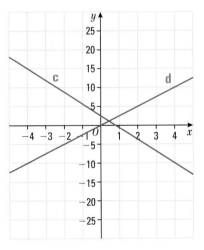

C 3 Using separate axes, using 1 cm to represent 1 unit, draw straight lines with:
 a gradient 1, y-intercept -1
 b gradient 2, y-intercept -2
 c gradient $\frac{1}{2}$, y-intercept 3
 d gradient -3, y-intercept 1
 e gradient $-\frac{1}{3}$, passing through the point $(0, 1)$.

C 4 A straight line has gradient 2. The point $(3, -1)$ lies on the line. Find the coordinates of one other point on the line.

5 P is the point $(-3, 5)$. Q is the point $(6, 0)$.
 a Find the gradient of the line PQ.
 b Find the coordinates of the y-intercept of the line PQ.

Exercise 15F

D 1 The graph shows the cooking time, t minutes, needed for a leg of lamb of weight w kilograms.

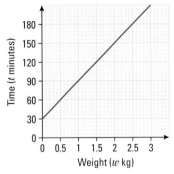

 a Work out the gradient of the graph and explain what it represents.
 b Describe a rule to give the time needed to cook a leg of lamb of any weight.
 c Why do you think the line doesn't extend down to the x-axis?

2 This graph can be used to change between distances measured in kilometres and distances measured in miles.

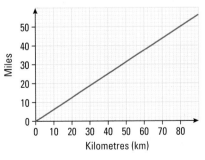

 a What distance in miles is equivalent to 40 km?
 b Find the gradient of this graph and explain what it represents.

C 3 The diagram shows the distance–time graphs of a marathon runner, a car and a train.

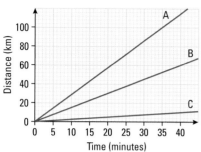

a Work out the gradient of each graph.

b Use your answer to part **a** to decide which mode of transport each of the graphs A, B and C could represent.
Give reasons for your answers.

4 Fiona is topping up the water in her pond. She records the amount of water in the pond every 15 seconds. Her results are shown in the table.

Time (t seconds)	Water (w litres)
0	75
15	83
30	91
45	99
60	107
75	115
90	123

a Draw a straight-line graph to show the amount of water for $t = 0$ to $t = 90$.

b Work out the gradient of the straight line.

c Describe what the gradient represents.

15.4 The equation $y = mx + c$

Exercise 15G

C 1 A line which passes through the point $(0, -3)$ has gradient 7.
Write down the equation of the line.

2 Find **i** the gradient and **ii** the y-intercept of the lines with the equations

a $y = 3x + 2$ b $y = 2x - 4$

c $y = \frac{2}{5}x - 3$ d $3x + 5y = 10$

e $3x - 4y = 12$ f $2x - y = 0$

B 3 Find the equations of the lines shown in the diagram.

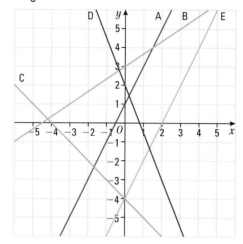

A 4 A line passes through the points with coordinates $(2, 4)$ and $(1, 6)$.
Find the equation of the line.

5 The gradient of a line is 2. The point with coordinates $(5, 3)$ lies on the line.
Find the equation of the line.

15.5 Parallel and perpendicular lines

Exercise 15H

B 1 Copy and complete the following table to show the gradients of pairs of lines l_1 and l_2 which are perpendicular to each other.

	a	b	c	d	e
Gradient of line l_1	2	−3	$\frac{1}{4}$		
Gradient of line l_2				−2	$\frac{1}{5}$

2 Write down the equation of a line parallel to the line with the equation

a $y = 3x + 9$ b $y = \frac{1}{4}x - 2$

c $y = 2 - x$

3 Write down the equation of a line perpendicular to the line with the equation

a $y = x + 4$ b $y = 3x - 1$

c $y = 2 - \frac{1}{5}x$

A 4 Find the equation of a line which is parallel to the line with the equation $y = 3x - 1$ and which passes through the point $(0, 6)$.

A

5 Find the equation of a line which is parallel to the line with the equation $3x + y = 7$ and which passes through the origin.

6 Find the equation of a line which is perpendicular to the line with the equation $y = \frac{1}{5}x$ and which passes through the point $(-4, 8)$.

7 Find the equation of a line which is perpendicular to the line with the equation $x + y = 8$ and which passes through the point $(-2, 10)$.

15.6 Real-life graphs

Exercise 15I

C

1 Jessica has a job delivering leaflets advertising a new pizza delivery service. She walks to the pizza place to collect the leaflets. Once there, Jessica spends some time sorting the leaflets and finding out exactly where to go. She starts delivering the leaflets and has two rests on the way home.

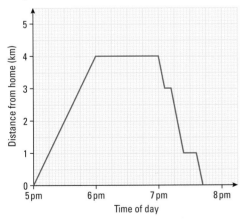

a How far away from Jessica's home is the pizza place?

b How long does it take Jessica to sort the leaflets?

c How fast does Jessica walk on the way to work?

d What is the total amount of rest time Jessica has while delivering the leaflets?

e What is the average speed of Jessica's delivery round?

C

2 Phillip carried out an experiment to investigate how quickly the water in a kettle cools down. He boiled the kettle then recorded the temperature of the water at regular time intervals.
He used his results to draw the graph shown.

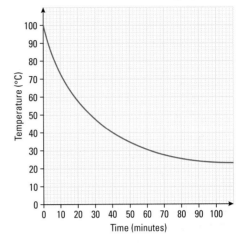

a What was the temperature of the water 15 minutes after the kettle had boiled?

b How long did it take the water to cool down to 30°C?

c Explain what the shape of the graph shows about the way in which the temperature of the water changes.

d Phillip said, 'The temperature of the water did not fall below 22°C.' Phillip was correct. Can you explain why this might be?

3 Liquid is poured into each of these containers. Sketch a graph to show the relationship between the depth of water and the volume of water in each container.

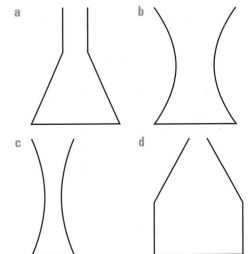

C **4**

A02
A03

Andrew is renovating a house. He wants to hire a floor sander. The graph shows some information about the cost of hiring a sander.

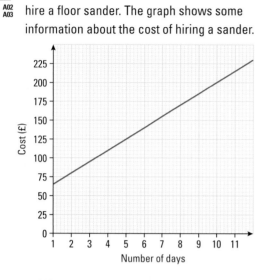

a Work out the gradient of the graph. What does this represent?

b How much does it cost to hire a sander for one day?

c Explain the rule for working out the cost of hiring the sander for several days.

C **5**

A03

David is buying a small van. He looks up the petrol consumption before making a decision.

Speed (km/h)	62	70	80	85	98	105
Petrol consumption (km/l)	12.4	13.4	14.2	14.3	13.6	12.8

Draw axes on graph paper, taking 2 cm to represent 10 km/h on the horizontal axis (start at 60 km/h) and 4 cm to represent 1 km/l on the horizontal axis (start at 12 km/l).

Plot the values from the table and join them with a smooth curve. Use your graph to estimate:

a the petrol consumption at 75 km/h

b the speeds that give a petrol consumption of 14 km/l.

16 Ratio and proportion

Key Points

- **ratio:** a comparison of a part to a part, written in the form $a : b$.
    - the simplest form of a ratio has whole numbers with no common factor
- **unitary form:** a ratio written in form $1 : n$. Often used for scales in maps.
- **equivalent ratios:** equal ratios like $15 : 20$ and $3 : 4$.
- **direct proportion:** when two quantities increase and decrease in the same ratio.
- **inverse proportion:** when one quantity increases at the same rate as the other quantity decreases.

- **writing a ratio in unitary form:** divide each number in the ratio by the first number in the ratio.
- **sharing in a given ratio:** use one of these methods
    - work out how much each share is worth and multiply by the number of shares each person receives
    - work out the fraction of the total each person receives and multiply the total by these fractions

16.1 Introducing ratio

Exercise 16A

Questions in this chapter are targeted at the grades indicated.

D **1** Write each ratio in the form $1 : n$.
   a $6 : 18$ b $7 : 35$ c $4 : 18$
   d $8 : 4$ e $36 : 9$ f $21 : 9$
   g $\frac{1}{2} : 5$ h $\frac{5}{8} : \frac{2}{10}$

2 In a pond, there are 40 moorhens. There are 160 ducks. Write down the ratio of the number of moorhens to the number of ducks.
Give your ratio in the form $1 : n$.

3 In a salsa club, there are 160 men and 220 women.
   a What fraction of the club members are men?
   b Write down the ratio of the number of men to the number of women.
      Give your ratio in its simplest form.
   c Write your answer to part **b** in the form $1 : n$.

4 A scale model of a building has a height of 9 cm. The height of the real building is 30 m.
Work out the ratio of the height of the model to the height of the real building.
Write your answer in the form $1 : n$.

D **5** Write these ratios in the form $1 : n$.
   a 2 hours : $\frac{1}{4}$ hour   b £6 : 80p
   c 3 m : 4 cm   d 40 g : 1 kg

16.2 Solving ratio problems

Exercise 16B

D **1** To make orange squash, the ratio of water to orange concentrate is $4 : 1$. Work out the amount of concentrate needed for:
   a 62 cl of water
   b 160 cl of water
   c 2.8 l of water.

2 Jean mixes green paint using blue and yellow in the ratio $2 : 5$.
   a If Jean uses 140 ml of yellow paint, work out how much blue paint she needs to add.
   b If she uses 60 ml of blue paint, how much yellow paint does she need?

3 A map is drawn using a scale of $1 : 250\,000$. On the map, the distance between two towns is 23.5 cm.
Work out the real distance between the towns. Give your answer in kilometres.

C **4** Luisa and Christina share some money in the ratio 7 : 8.
If Luisa receives £420, work out how much money Christina gets.

5 The ratio of the lengths of two tables is 6 : 8.
If the length of the first table is 1.26 m, calculate the length of the second table.

6 In a ball pool, the ratio of red balls to blue balls is $1 : \frac{3}{5}$. If there are 120 blue balls in the pool, work out the number of red balls.

16.3 Sharing a quantity in a given ratio

Exercise 16C

C **1** Divide the quantities in the ratios given.
a £21.91 in the ratio 3 : 4
b 550 cm in the ratio 3 : 2
c £290.64 in the ratio 1 : 4 : 7
d 75.24 kg in the ratio 3 : 7 : 8

2 The angles in a triangle are in the ratio 15 : 10 : 11.
Find the sizes of the three angles.

3 A02 A03 Three friends got some work stuffing envelopes. They were paid £83.20 altogether.
They shared the money in the ratio of the amount of time that each of them worked. Susie worked for 5.5 hours, Xiu worked for 4.5 hours and Rob worked for 3 hours.
Calculate the amount of money Xiu received.

4 A03 Sue and Simon won £280 in a raffle. They share the prize in the ratio of 4 : 3. Sarah forgot to buy a ticket, so Sue gives her a quarter of her share and Simon gives her a third of his share.
What fraction of the prize does Sarah receive? Give the fraction in its lowest terms.

5 A03 A farm makes up boxes of vegetables.
They contain carrots, onions and leeks in the ratio of 6 : 5 : 4. There are always more than 40 vegetables in a box.
What is the minimum number of onions in a box?

C **6** A03 Naomi, Fiona and Miriam are going on a driving holiday. They will share the cost of the petrol in the ratio 5 : 6 : 7. Naomi fills the car with petrol and pays her share of £12.50.
a How much did it cost to fill the car with petrol?
b What fraction of the total cost was Fiona's share?

16.4 Using direct proportion

Exercise 16D

1 A lorry travels at a steady speed of 70 km each hour.
Work out the number of hours it takes to travel:
a 350 km
b 420 km.

2 Five 2 GB memory cards cost a total of £61.50.
Work out the cost of eight of the 2 GB memory cards.

D **3** A03 Sandra is paid £61.20 for 9 hours' work in a warehouse.
How much is she paid for 4 hours' work?

4 Jane buys 14 rulers for £11.90.
Work out the cost of 9 of these rulers.

B **5** A03 The cost of fabric is directly proportional to its length. A 4.5 m piece of fabric costs £2.88.
Work out the cost of 7 m of this fabric.

A **6** A03 When a printer produces photo enlargements, he makes sure that the width is directly proportional to the length so as not to stretch the image. He enlarges one photo so it has width 0.5 m and length 0.7 m. He then prints another enlargement for a mural with length 2.45 m.
What is the width of the mural?

Exercise 16E

1 A02 A03 These are the ingredients for 12 muffins.
70 ml milk   1 egg
40 ml sunflower oil   80 g sugar
300 g flour   100 g blueberries
a Work out the amount of sugar needed to make 36 muffins.
b Work out the amount of milk needed to make 30 muffins.

2 A machine produces 780 doughnuts in 1 hour. How long will the machine take to produce 4485 doughnuts?
Give your answer in hours and minutes.

D **3** The exchange rate is £1 = $1.40.
A03
  a Convert £300 to dollars.
  b Convert $658 to pounds.

4 Carl bought a painting for €168 in Italy.
A03 The exchange rate was £1 = €1.20.
Work out the cost of the painting in pounds.

5 * Michael buys a camera in London for £180.
A03 He then travels to New York. He sees the identical camera on sale for $200.
If the exchange rate is £1 = $1.51, in which city is the camera cheaper and by how much?

6 * Gillian travelled to Paris in December.
A03 She exchanged £350 for euros at a rate of £1 = €1.108. She is going to Paris again in March and is exchanging the same number of pounds. In March, the exchange rate is £1 = €1.118. How many more euros will Gillian receive in March than she did in December?

16.5 Using inverse proportion

Exercise 16F

D **1** It took 12 electricians 3 days to rewire a building.
A03 Work out how long it would have taken to rewire the building if there had been:
  a 6 electricians
  b 5 electricians.

2 If 10 machines can weave a certain amount of fabric in 5 hours, how long will 15 machines take to weave the same amount of fabric?
A02
A03

3 It takes 12 technicians 4 hours to enter data into a computer. How long will it take 8 technicians to enter the data?
A02
A03

4 A school is cleaned in 6 hours by 3 cleaners. How long will it take 4 cleaners to clean the school?
A02
A03

5 In one batch, a factory produces 450 tiles measuring 15 cm long by 12 cm wide. Using the same raw materials, it produces another batch of 450 tiles, this time with a length of 20 cm. What is the width of these tiles?
A02
A03

6 A document will fit onto exactly 36 pages if there are 500 words on a page.
A02
A03 If the number of words on each page is reduced to 300, how many more pages will there be in the document?

17 Transformations

⊚ **translation:** a sliding movement where lengths and angles do not change, and the shape does not turn.

  ⊚ **column vector:** a way of describing a translation, e.g. $\begin{pmatrix} 3 \\ 6 \end{pmatrix}$. The top number describes the movement to the right parallel to the x-axis, and the bottom number describes the movement up parallel to the y-axis.

⊚ **reflection:** an image of a shape produced by reflecting an object in the line of reflection or mirror line. Lengths and angles do not change.

⊚ **rotation:** turning by a fraction of a turn or by an angle. Lengths and angles do not change.

  ⊚ **centre of rotation:** the point about which the shape is turned

⊚ **enlargement:** a change in the size of a shape. Lengths change but the angles do not.

  ⊚ **scale factor:** the value that multiplies the lengths to give the enlarged image
  scale factor = length of side in image ÷ length of corresponding side in original object

  ⊚ **centre of enlargement:** the point from which the enlarged lengths are measured from.

17.1 Using translations

⚙ **Exercise 17A**

Questions in this chapter are targeted at the grades indicated.

C **1** Describe, with a vector, the translation that maps triangle A onto:

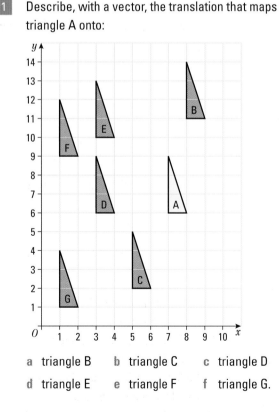

  a triangle B **b** triangle C **c** triangle D

  d triangle E **e** triangle F **f** triangle G.

C **2** On a copy of the diagram translate triangle A:

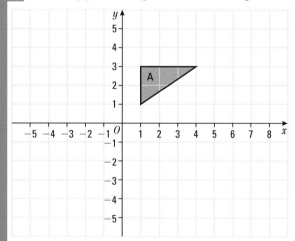

  a 4 to the right and 2 up.
  Label your new triangle B.

  b 5 to the left and 4 down.
  Label your new triangle C.

  c 6 to the left. Label your new triangle D.

  d by the vector $\begin{pmatrix} 2 \\ -3 \end{pmatrix}$.
  Label your new triangle E.

  e by the vector $\begin{pmatrix} -4 \\ -6 \end{pmatrix}$.
  Label your new triangle F.

C 3 The coordinates of point A of this kite are $(-5, 5)$. The kite is translated so that the point A is mapped onto the point $(3, 2)$.

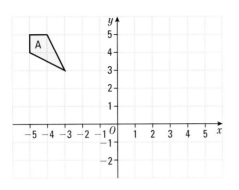

a On a copy of the diagram draw the image of the kite after this translation.

b Describe this translation with a vector.

4 Draw the following translations on a copy of the diagram.

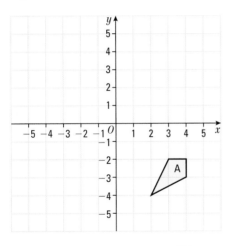

a Translate kite A by the vector $\begin{pmatrix} 1 \\ 6 \end{pmatrix}$.
Label this new kite B.

b Translate kite B by the vector $\begin{pmatrix} -5 \\ -2 \end{pmatrix}$.
Label this new kite C.

c Describe, with a vector, the translation that maps kite A onto kite C.

d Describe, with a vector, the translation that maps kite C onto kite A.

17.2 Transforming shapes using reflections

Exercise 17B

D 1
A03 Make a copy of the diagram and complete the following reflections.

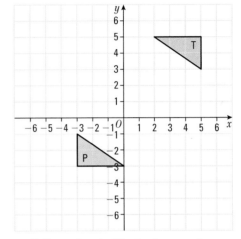

a Reflect triangle P in the line $x = 1$.
Label this new triangle Q.

b Reflect triangle P in the line $y = 1$.
Label this new triangle R.

c Describe the reflection that maps triangle Q onto triangle T.

C 2
A03 On a copy of the diagram, complete the following reflections.

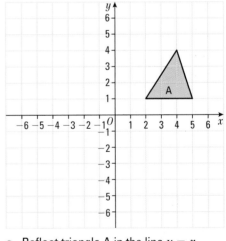

a Reflect triangle A in the line $y = x$.
Label this new triangle B.

b i Reflect triangle A in the line $y = -x$.
Label this new triangle C.

   ii Reflect triangle C in the line $y = x$.
Label this new triangle D.

c Describe fully the transformation that maps triangle B onto triangle D.

C 3

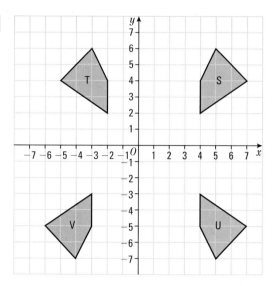

a Give the equation of the mirror line of the reflection that maps:
  i shape S onto shape T
  ii shape V onto shape U.
b Describe fully the transformation that maps shape U onto shape S.

17.3 Transforming shapes using rotations

Exercise 17C

D 1 On a copy of the diagram, complete the following rotations.

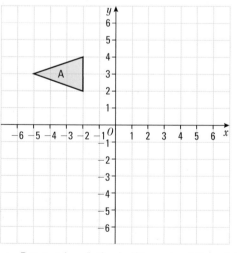

a Rotate triangle A a half turn about the origin O. Label the new triangle B.
b Rotate triangle A a quarter turn clockwise about the origin O. Label the new triangle C.
c Rotate triangle A a quarter turn anticlockwise about the origin O. Label the new triangle D.

C 2 Make three copies of this diagram showing trapezium P.

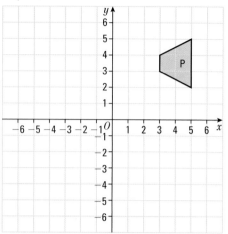

a On copy 1 of the diagram, rotate trapezium P 180° about the point (0, 2).
  Label the new trapezium Q.
b On copy 2 of the diagram, rotate trapezium P 90° clockwise about the point (1, 2).
  Label the new trapezium R.
c On copy 3 of the diagram, rotate trapezium P 90° anticlockwise about the point (2, −3).
  Label the new trapezium S.

3
A03

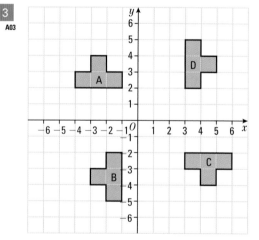

a Describe fully the rotation that maps shape A onto: i shape B ii shape C iii shape D.
b Describe fully the rotation that maps shape B onto shape A.
c Describe fully the rotation that maps shape D onto shape B.

ResultsPlus
Exam Tip

A common mistake when describing a rotation is to call it a turn instead of a rotation and forgetting to say where the centre of rotation is.

C 4

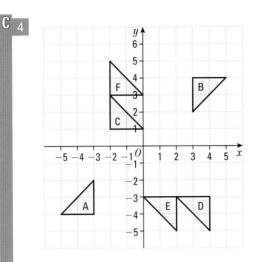

a Describe fully the rotation that maps triangle A onto:
  i triangle B
  ii triangle C
  iii triangle D
  iv triangle E
  v triangle F.

A03 b Describe the transformation that maps triangle C onto triangle D.

c Describe the transformation that maps:
  i triangle D onto triangle A
  ii triangle F onto triangle E.

17.4 Enlargements and scale factors

 Exercise 17D

1 Here is a right-angled triangle.

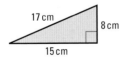

The triangle is enlarged with a scale factor of 3.
a Work out the length of each side of the enlarged triangle.
b Compare the perimeter of the enlarged triangle with the perimeter of the original triangle.

2 Copy the shape on squared paper and draw:

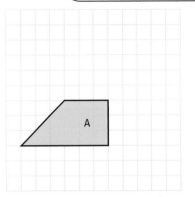

a an enlargement of shape A with scale factor 2. Label this enlargement shape B.

b an enlargement of shape A with scale factor $\frac{1}{3}$. Label this enlargement shape C.

A02
A03 c Shape B is an enlargement of shape C. Work out the scale factor of the enlargement.

D 3

A03 Rectangle S has a base of 5 cm and a height of 3 cm.
Rectangle T is an enlargement of rectangle S with a scale factor of 2.
Rectangle U is an enlargement of rectangle S with a scale factor of 3.
a On squared paper, draw rectangles S, T and U.
b Find the perimeter of:
  i rectangle S ii rectangle T iii rectangle U.
c Find the area of:
  i rectangle S ii rectangle T iii rectangle U.
d Work out the value of:
  i $\dfrac{\text{Perimeter of T}}{\text{Perimeter of S}}$ ii $\dfrac{\text{Perimeter of U}}{\text{Perimeter of S}}$.
  Write down anything that you notice about these values.
e Work out the value of:
  i $\dfrac{\text{Area of T}}{\text{Area of S}}$ ii $\dfrac{\text{Area of U}}{\text{Area of S}}$.
  Write down anything that you notice about these values.
f Rectangle V is an enlargement of rectangle S with a scale factor of 6.
  What is the perimeter of rectangle V?

Exercise 17E

1 Copy the shape on squared paper and draw the enlargement of the shape with the given scale factor and centre of enlargement marked with a dot (•).

a Scale factor 2.

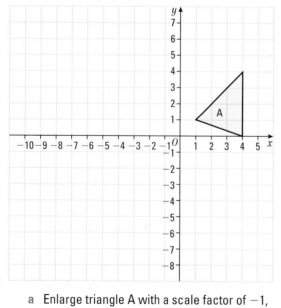

b i Scale factor 2.
ii Scale factor 0.5.
Draw both enlargements on the same diagram.

2 On a copy of the diagram complete the following enlargements.

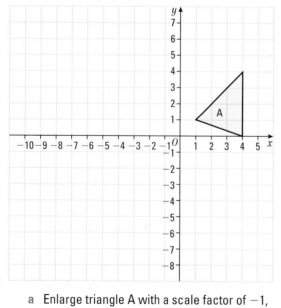

a Enlarge triangle A with a scale factor of -1, centre $(-2, 0)$. Label this new triangle B.

b Enlarge triangle A with a scale factor of $-\frac{1}{3}$, centre $(-1, 4)$. Label this new triangle C.

A03 **c** Find the scale factor of the enlargement that maps triangle C onto triangle B.

3

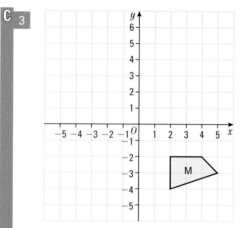

a On a copy of the diagram, enlarge shape M with a scale factor of -1, centre $(1, 1)$. Label this new shape N.

The mapping of shape M onto shape N is also a rotation.

A03 **b** Describe fully the rotation that maps shape M onto shape N.

17.5 Combinations of transformations

Exercise 17F

For each question, make a copy of the diagram.

1 Complete the following translations.

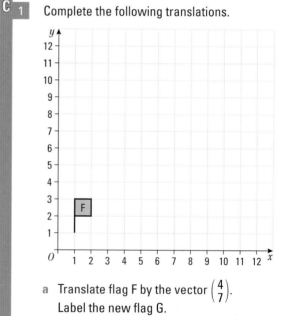

a Translate flag F by the vector $\begin{pmatrix} 4 \\ 7 \end{pmatrix}$.
Label the new flag G.

b Translate flag G by the vector $\begin{pmatrix} 5 \\ -3 \end{pmatrix}$.
Label the new flag H.

A03 **c** Describe fully the single transformation which maps flag F onto flag H.

A03 **d** Describe fully the single transformation which maps flag H onto flag F.

2 Complete the following transformations.

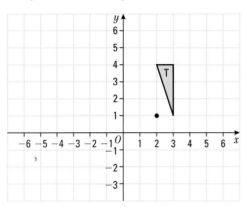

a Rotate triangle T 180° about (2, 1).
   Label the new triangle U.

b Translate triangle U by the vector $\begin{pmatrix} -3 \\ 3 \end{pmatrix}$.
   Label the new triangle V.

A03 c Describe fully the single transformation which
      maps triangle T onto triangle V.

3 Complete the following transformations.

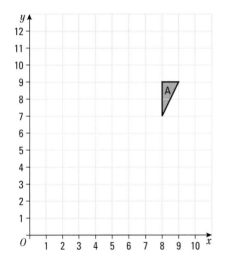

a Rotate triangle A 90° clockwise about (6, 7).
   Label the new triangle B.

b Rotate triangle B 90° clockwise about (3, 6).
   Label the new triangle C.

c Rotate triangle B 90° anticlockwise about
   (3, 6).
   Label the new triangle D.

A03 d Describe fully the single transformation that
      maps triangle A onto:
      i triangle C
      ii triangle D.

4 Complete the following transformations.

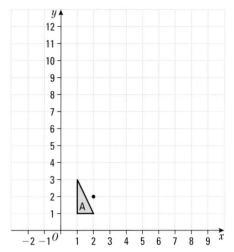

a Enlarge triangle A with scale factor 4 and
   centre of enlargement O.
   Label the new triangle B.

b Enlarge triangle B with scale factor $\frac{1}{2}$ and
   centre of enlargement (2, 2).
   Label the new triangle C.

A03 c Describe fully the single transformation that
      maps triangle A onto triangle C.

5 Complete the following transformations.

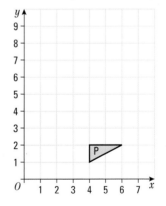

a Reflect triangle P in the line $y = 3$.
   Label the new triangle Q.

b Reflect triangle Q in the line $y = 6$.
   Label the new triangle R.

A03 c Describe fully the single transformation that
      has the same effect as a reflection in $y = 3$
      followed by a reflection in $y = 6$.

18 Processing, representing and interpreting data

Key Points

- **piechart:** a circle divided into sectors to show how the total is split up between different categories.
  - area of each sector represents the number of items in that category
  - sector angle $= \dfrac{\text{frequency} \times 360°}{\text{total frequency}}$
  - frequency $= \dfrac{\text{sector angle} \times \text{total frequency}}{360°}$
  - angles at the centre of the pie chart add up to 360°
- **stem and leaf diagram:** a way of presenting data to show its pattern and make it easier to find the quartiles.
- **composite bar chart:** a chart that shows the size of individual categories split into separate parts.
- **comparative or dual bar chart:** a chart that shows two or more bars drawn side by side for each category.
- **frequency diagram:** a chart used to represent grouped discrete data.

- **histogram:** a graph used to display grouped continuous data.
- **cumulative frequency of a value:** the total number of observations that are less than or equal to that value.
  - a cumulative frequency graph can be used to estimate the quartiles
- **box plots:** diagrams that show the median, upper and lower quartiles and the maximum and minimum values. Often used to compare distributions.
- **drawing a frequency polygon for continous data:** draw a histogram for the data, mark the midpoints of the tops of the bars and join them with straight lines.
- **drawing histograms with unequal class intervals:** adjust the height of the bars by using a scale of frequency density, where:

$$\text{frequency density} = \dfrac{\text{frequency}}{\text{class width}}$$

  - the area of each bar gives its frequency

18.1 Producing pie charts

Exercise 18A

Questions in this chapter are targeted at the grades indicated.

1 * The numbers of teas and coffees sold in a canteen one lunchtime are shown in the table.

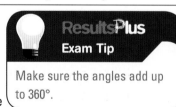

ResultsPlus
Exam Tip

Make sure the angles add up to 360°.

Type of drink	Number of drinks
Breakfast tea	48
Peppermint tea	20
Fruit tea	22
Espresso	20
Latte	42
Cappuccino	28

Draw a pie chart to represent these data.
Use a radius of 3.5 cm.

2 * A newsagent sold 90 packets of crisps in one hour.

The table shows the number of each flavour of crisps sold.

Flavour of crisps	Number of packets
Cheese and onion	18
Salt and vinegar	16
Barbecue beef	33
Smokey bacon	17
No added salt	6

Draw a pie chart to represent these data.
Use a radius of 3.5 cm.

3* A store manager in London asks his employees how they travel to work.
The table shows these data.

Method of getting to work	Number of employees
Bus	45
Car	6
Cycle	4
Underground train	48
Overground train	15
Motorbike	2

Draw a pie chart to represent these data.

18.2 Interpreting pie charts

Exercise 18B

1 The pie chart shows how the 240 girls in Year 10 at Cotly High School chose from five lunch club activities.

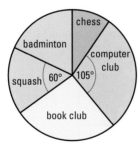

a Write down the least popular option.
b Write down the most popular option.
c Work out how many girls chose squash.
d Work out how many girls chose computer club.

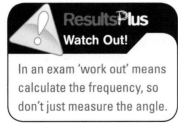

ResultsPlus
Watch Out!

In an exam 'work out' means calculate the frequency, so don't just measure the angle.

2* A company owns two sandwich shops in Fourdon.
They do a survey to find the number of each type of sandwich they sell at lunchtime on one particular day.

Sandwich type	Frequency	
	Shop S	Shop T
Cheese	18	24
Tuna	22	26
Prawn	20	18
Ham	5	10
Chicken	7	12

Compare and contrast the information by drawing two pie charts.

18.3 Representing and interpreting data in a stem and leaf diagram

Exercise 18C

C 1 Noah records the number of hits on his website he receives every day for 40 days.
The data he collects are shown in the stem and leaf diagram.

0	1	2	2	5	6			
1	2	2	3	4	5	5	7	
2	3	3	4	5	6	7	8	9
3	3	3	4	4	5	5	7	8	8
4	2	3	3	3	4	6	6	
5	0	1	1	3				

Key 3 | 4 stands for 34

a Write down the mode of these data.
b Find the median of these data.
c Work out the range of these data.
d Find Q_1 and Q_3 of these data.
e Work out the interquartile range for these data.

C **2** A dentist recorded how many minutes late his patients were for their appointments over a period of 20 working days.

| 5 | 7 | 7 | 8 | 5 | 10 | 20 | 10 | 15 | 14 |
| 10 | 8 | 5 | 15 | 18 | 5 | 5 | 14 | 16 | 17 |

a Draw an ordered stem and leaf diagram for these data.

b Use your stem and leaf diagram to find the mode of these data.

c Use your stem and leaf diagram to find the median of these data.

d Work out the range of these data.

e Use your stem and leaf diagram to find Q_1 and Q_3 of these data.

f Work out the interquartile range for these data.

3 Fiona manages several pharmacies. She has to drive between them. These are the lengths of her journeys on 20 days last month.
The distances are in kilometres.

| 34 | 21 | 6 | 14 | 23 | 35 | 20 | 30 | 23 | 35 |
| 39 | 8 | 20 | 12 | 21 | 17 | 12 | 23 | 26 | 6 |

a Draw an ordered stem and leaf diagram for these data.

b Use your stem and leaf diagram to find the mode of these data.

c Use your stem and leaf diagram to find the median of these data.

d Work out the range of these data.

e Use your stem and leaf diagram to find Q_1 and Q_3 of these data.

f Work out the interquartile range for these data.

18.4 Interpreting comparative and composite bar charts

Exercise 18D

1 The composite bar chart shows the rainfall in a number of cities in March and November.

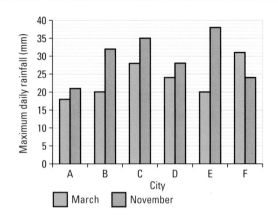

a Write down the highest maximum daily rainfall in March.

b Write down the highest maximum daily rainfall in November.

c Write down the cities that had the same maximum daily rainfall in March.

d Write down the city in which the maximum daily rainfall in November was 32 mm.

e Write down the city in which the maximum daily rainfall in March was 28 mm.

2 The composite bar chart shows how Jen spends her wages.

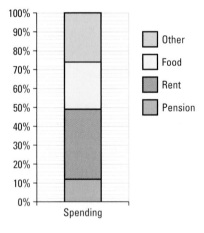

a What did Jen spend most on?

b What did Jen spend least on?

c What percentage of her wages does Jen spend on 'other' items?

3 The composite bar charts show the make-up of 100 grams of each of two cans of soup.

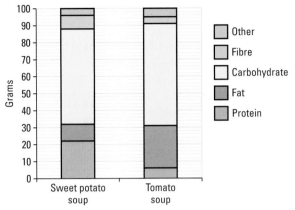

a How many grams of fibre are there in 100 g of sweet potato soup?

b Estimate the number of grams of fat in 100 g of tomato soup.

c Write down the type of soup that has more carbohydrates.

4 Mr Yu wants to compare three popular dishes in his restaurant. The composite bar chart shows the sales of each dish on two consecutive evenings.

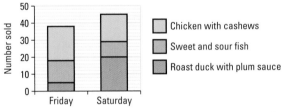

a On which day were most meals sold overall?

b On which day were most fish dishes sold?

c How many chicken dishes were sold on Saturday?

18.5 Drawing and interpreting frequency diagrams and histograms

Exercise 18E

D 1 The grouped frequency table shows information about the number of music tracks downloaded by each of 30 students in one week.

Draw a frequency diagram for this information.

Number of tracks	Frequency
0 to 2	4
3 to 5	5
6 to 8	9
9 to 11	8
12 to 14	4

D 2 The grouped frequency table shows information about the weights of 35 onions.

Weight (w g)	Frequency
$155 \leqslant w < 160$	8
$160 \leqslant w < 165$	15
$165 \leqslant w < 170$	7
$170 \leqslant w < 175$	3
$175 \leqslant w < 180$	2

a Write down the modal class.

b The last onion weighed was 175 g. In which class interval is this recorded?

c Draw a histogram for these data.

3 A farmer weighed his 48 piglets. Some information about their weights is shown in the table.

Weight (w pounds)	Frequency
$40 \leqslant w < 55$	4
$55 \leqslant w < 70$	12
$70 \leqslant w < 85$	16
$85 \leqslant w < 100$	10
$100 \leqslant w < 115$	6

a Write down the modal class.

b In which class interval does the weight of 90 pounds fall?

c Draw a histogram for these data.

18.6 Drawing and using frequency polygons

Exercise 18F

C 1 A factory boss wants to find out how many red sweets there are in each pack of wine gums her factory produces. She records the number of red sweets in one batch of 50 packs.

The table shows this information.

Number of red sweets	3	4	5	6	7	8
Frequency	2	3	4	16	15	10

Draw a frequency polygon for these data.

C 2 The noise levels at 35 nightclubs were measured in decibels.

The data collected are shown in the grouped frequency table.

Noise level (d decibels)	Frequency
$80 \leqslant d < 90$	5
$90 \leqslant d < 100$	15
$100 \leqslant d < 110$	11
$110 \leqslant d < 120$	4

a Write down the modal class.

b Use the information in the table to draw a histogram.

c Use your answer to part **b** to draw a frequency polygon.

3 A competition is held for the longest cucumbers grown in local allotments. The lengths, in centimetres, of all the cucumbers were measured.

The information collected is shown in the table.

Cucumber length (l cm)	Frequency
$25 \leqslant l < 27$	3
$27 \leqslant l < 29$	5
$29 \leqslant l < 31$	7
$31 \leqslant l < 33$	8
$33 \leqslant l < 35$	7

Draw a frequency polygon for these data.

4 * The two frequency polygons show the amount of
A03 time it took a group of teenagers and a group of adults to do a sudoku puzzle.

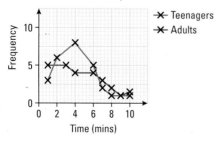

Who were better at doing the puzzle, teenagers or adults?

Give a reason for your answer.

18.7 Drawing and using histograms with unequal class intervals

Exercise 18G

A 1 The table gives information about the lifetime of a new energy-saving light bulb.

Lifetime t hours (1000s)	Frequency	Class width	Frequency density
$0 \leqslant t < 5$	2		
$5 \leqslant t < 8$	8		
$8 \leqslant t < 10$	16		
$10 \leqslant t < 12$	24		
$12 \leqslant t < 15$	15		

a Copy and complete the table.

b Draw a histogram for these data.

2 * The table gives information about the
A02 distances travelled by a group of MPs to their
A03 constituencies.

Distance (d kilometres)	Frequency
$0 < d \leqslant 10$	7
$10 < d \leqslant 15$	11
$15 < d \leqslant 20$	13
$20 < d \leqslant 30$	27
$30 < d \leqslant 40$	42

Draw a histogram for these data and find an estimate of the number of MPs who travel a distance of 10–30 km.

A☆ 3 The table gives information about the age of members of the audience at a pantomime performance.

A02 a Copy and complete the table and histogram, including the frequency density scale.

A☆

Age (a years)	Frequency
$0 < a \leqslant 5$	25
$5 < a \leqslant 10$	46
$10 < a \leqslant 15$	
$15 < a \leqslant 25$	
$25 < a \leqslant 40$	

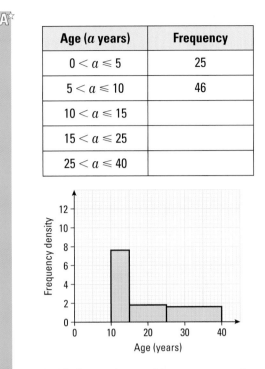

A03 **b** Find an estimate of how many members of the audience were aged between 15 years and 30 years.

18.8 Drawing and using cumulative frequency graphs

Exercise 18H

B **1** The table shows the ages of people using a swim and surf centre.

Age (x years)	Frequency	Cumulative frequency
$x \leqslant 10$	8	
$10 < x \leqslant 15$	18	
$15 < x \leqslant 20$	15	
$20 < x \leqslant 25$	10	
$25 < x \leqslant 30$	6	
$30 < x \leqslant 35$	2	
$35 < x \leqslant 40$	1	

a Copy and complete the table.
b Draw a cumulative frequency graph for these data.

B **2** The cumulative frequency graph shows the time a group of students spent on learning to drive.

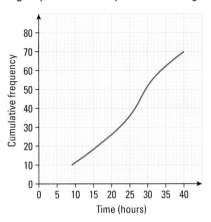

a Use the cumulative frequency graph to estimate the number of students who spent up to 25 hours learning to drive.
b Use the cumulative frequency graph to estimate the number of students who spent more than 30 hours learning to drive.
c Use the cumulative frequency graph to estimate the number of students who spent between $17\frac{1}{2}$ and $32\frac{1}{2}$ hours learning to drive.

3 The cumulative frequency graph shows the speeds of vans recorded by a camera on a stretch of dual carriageway.

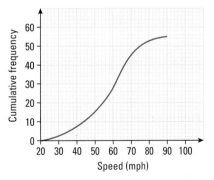

a Use the graph to find an estimate for the number of van drivers
  i driving at 40 mph or less
  ii driving between 30 mph and 50 mph.
b How many vans were recorded by the speed camera?
A03 **c** The speed limit for a van on a dual carriageway is 60 mph. Estimate the percentage of vans breaking the speed limit.

18.9 Finding quartiles from a cumulative frequency graph

Exercise 18I

B **1** The cumulative frequency graph shows the marks of 50 students in a music examination.

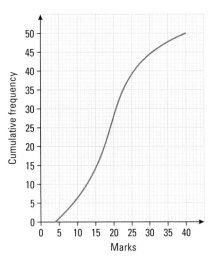

Marks

a Find an estimate for the median (Q_2).

b Find an estimate for Q_1 and Q_3.

c Work out the interquartile range.

d Work out the range.

2 The cumulative frequency graph shows the average hours of television a sample of people watch per day.

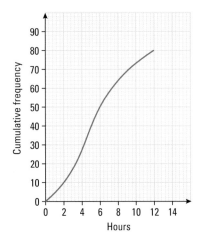

Hours

a Find an estimate for the median (Q_2).

b Find an estimate for Q_1 and Q_3.

c Work out the interquartile range.

B **3** The cumulative frequency graph shows the prices of watches in a jeweller's catalogue.

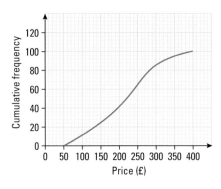

Price (£)

a Find estimates for the median and quartiles.

b Find estimates for the range and the interquartile range.

18.10 Drawing and interpreting box plots

Exercise 18J

B **1** A zoo keeper estimated the lengths of all the lizards in the reptile house. The longest was 30 cm and the shortest was 5 cm. The median length was 14 cm, the lower quartile 6.8 cm and the upper quartile 21.6 cm. Draw a box plot for these data.

2 The heights of delphiniums in a garden centre were measured in centimetres. They were as follows.

| 20 | 20.1 | 23.2 | 23.8 | 24 | 24.6 |
| 25 | 25.3 | 26 | 27.6 | 28 | 29.2 | 30 |

Draw a box plot for these data.

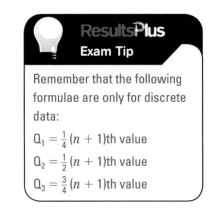

ResultsPlus

Exam Tip

Remember that the following formulae are only for discrete data:

$Q_1 = \frac{1}{4}(n + 1)$th value

$Q_2 = \frac{1}{2}(n + 1)$th value

$Q_3 = \frac{3}{4}(n + 1)$th value

3   The cumulative frequency graph gives information
     about the ages of turtles and parrots in a wildlife project.

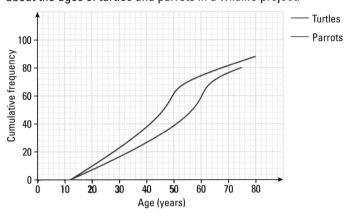

a   Use the cumulative frequency diagram to find the
     quartiles and the maximum and minimum values.

A03 b   Draw two box plots on the same scale using these
          data and compare and contrast the data.

19 Inequalities and formulae

Key Points

⊚ **inequality signs:**
- ⊚ $>$ means greater than
- ⊚ $\geqslant$ means greater than or equal to
- ⊚ $<$ means less than
- ⊚ $\leqslant$ means less than or equal to

⊚ **formula:** a way of describing a relationship between two or more sets of values.
- ⊚ **word formula:** where words represent the relationship between quantities
- ⊚ **algebraic formula:** where variables show a relationship between quantities, e.g. $E = mc^2$

⊚ **subject of a formula:** the variable that appears on its own on one side of the $=$ sign.

⊚ **drawing inequalities on a number line:** use an empty circle if the value is *not* included, and use a filled circle for a value that is included.

⊚ **solving linear inequalities:** use the same method as linear equations.
- ⊚ if you multiply or divide both sides by a negative number reverse the inequality sign
- ⊚ if there is an upper and lower limit, split the inequality into two and solve them separately

⊚ **solving inequalities graphically:** draw the lines on a graph and shade the regions that satisfy the inequality.
- ⊚ lines that are included in the region are draw with a solid line
- ⊚ lines that are not included are shown as a dotted line

⊚ **changing the subject of a formula:** carry out the same operations on both sides of the equal sign to isolate the subject.

19.1 Representing inequalities on a number line

Exercise 19A

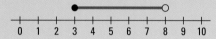
Questions in this chapter are targeted at the grades indicated.

C 1 Show these inequalities on a number line.

a $x \geqslant -4$

b $x < 0$

c $x < 4$ and $x > -3$

d $x \leqslant 2$ and $x \geqslant -2$

e $x < 3$ and $x \geqslant -4$

2 Write down the inequalities shown on these number lines.

a

b

c

d

C e

f

19.2 Solving simple linear inequalities in one variable

Exercise 19B

C 1 Solve these inequalities and show each answer on a number line.

a $x + 2 > 5$ b $x - 4 \leqslant -7$

c $2x + 7 \geqslant 1$ d $5x - 8 > 4$

B 2 Solve these inequalities.

a $4x < x + 12$ b $5x - 4 > 2x + 8$

c $2(x + 5) \leqslant 12$ d $5x - 8 > 2(x + 2)$

A★ 3 Solve these inequalities.

a $x + 2 \geqslant 5(x - 3)$ b $4(x - 1) < 3(x + 6)$

c $\dfrac{2 - 3x}{3} \leqslant \dfrac{x}{5} + 1$ d $\dfrac{4x + 3}{6} \geqslant 1 + \dfrac{2 + 3x}{4}$

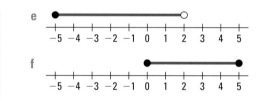

19.3 Finding integer solutions to inequalities in one variable

Exercise 19C

Find the possible integer values of x in these inequalities.

C 1
  a $-3 < x \leqslant 5$
       b $-6 < x < 3$
  c $0 \leqslant x < 4$
       d $-4 \leqslant x \leqslant 3$

2
  a $-6 < 2x \leqslant 8$
       b $-20 < 5x < 16$
  c $-7 \leqslant 10x < 40$
       d $-13 \leqslant 3x \leqslant 25$

B 3
  a $-3 \leqslant 2x + 3 < 7$
       b $-9 < 3x - 2 \leqslant 7$
  c $-20 \leqslant 5x - 7 < 12$
       d $-7 \leqslant 2x + 5 \leqslant 9$

A 4
  a $-2 < \dfrac{x}{3} \leqslant 1$
       b $-3 < \dfrac{3x}{4} \leqslant 2$

  c $-2 \leqslant \dfrac{3x - 2}{4} < 1$
       d $-2 < \dfrac{3 - 2x}{5} \leqslant 3$

19.4 Solving graphically several linear inequalities in two variables

Exercise 19D

D 1 x and y are integers.
$$-1 < x \leqslant 2 \qquad y > -1 \qquad y < x + 2$$
On a copy of the grid, mark with a cross (x), each of the nine points which satisfy all of these three inequalities.

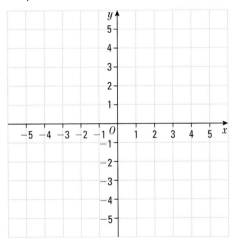

B 2   On a grid scaled from -6 to 6 on each axis, shade the region of points whose coordinates satisfy the following inequalities.
  a $x < 3$
       b $y \leqslant 2$
  c $-3 < x \leqslant 4$
       d $-3 \leqslant y \leqslant 1$

B 3   On a grid scaled from -6 to 6 on each axis, shade the region of points whose coordinates satisfy the following inequalities.
  a $-2 < x \leqslant 3$ and $-4 \leqslant y < 1$
  b $-2 \leqslant x \leqslant 1$ and $-2 < y \leqslant 4$
  c $-1 \leqslant x < 2$, $y > -3$ and $y \leqslant x$
  d $x > 0$, $y + x \leqslant 3$ and $y > 3x - 1$

A 4   The diagram shows a shaded region bounded by three lines.

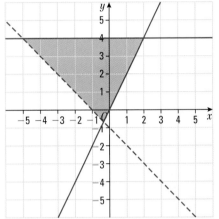

  a   Write down the equation of each of these lines.
  b   Write down the three inequalities satisfied by the coordinates of the points in this shaded region.
  c   If x and y are integers, write down the coordinates of the points in this shaded region.

5   The diagram shows a shaded region bounded by four lines.

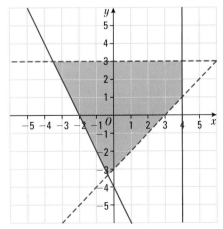

  a   Write down the equation of each of these four lines.
  b   Write down the four inequalities satisfied by the coordinates of the points in this shaded region.
  c   If x and y are integers, write down the least value of y in this shaded region.

19.5 Using formulae

Exercise 19E

> **ResultsPlus**
> **Exam Tip**
>
> The first step to finding a value from a formula is to simply replace each word or letter by its value.

1 $v = u + at$

Work out the value of v when

   a $u = 60$, $a = 8$ and $t = 3$

   b $u = 45$, $a = -3$ and $t = 10$

2 $N = 4m^3 - m^2$

Work out the value of N when

   a $m = 6$ **b** $m = -2$

3 $A = 2\pi r^2 + 2\pi rh$

Work out the value of A when

   a $\pi = 3.14$, $r = 15$ and $h = 20$

   b $\pi = 3.14$, $r = 4.2$ and $h = 10$

4   Use the formula **distance = speed × time** to work out:

   a   the distance travelled by a car travelling for $2\frac{1}{4}$ hours at an average speed of 58 mph

   b   the average speed of an athlete running 400 metres in 64 seconds.

5   The cooking time, in minutes, of a beef joint is given by the following formula:

   cooking time = weight of beef in kg × 35 + 30

   Use this formula to work out:

   a   the cooking time for a beef joint weighing 4 kg, giving your answer in hours and minutes

   b   the weight, in kg, of a beef joint taking 4 hours to cook.

6 $a = \sqrt{b^2 + c^2}$

Work out the value of a when

   a $b = 12$ and $c = 9$ **b** $b = 35$ and $c = 60$

19.6 Deriving an algebraic formula

Exercise 19F

1   A plumber charges £60 an hour, plus a £40 callout fee.

   a   Write a formula that can be used to find the total cost, £C, for a job that takes the plumber h hours.

   b   Use your formula to work out the cost of calling out the plumber for a job that takes 7.5 hours.

2   In a games tournament, 3 points are awarded for a win, 2 points are awarded for a draw and 1 point is deducted for a loss.

Denisha wins r games, draws s games and loses t games.

Write down a formula, in terms of r, s and t, for the total points (P) scored by Denisha.

3   Vicky works in a shop. She earns £v an hour when she works on a weekday and £w an hour when she works on a weekend. One week, she works 5 hours on each of Tuesday, Wednesday and Thursday, as well as 6 hours on Saturday and 4 hours on Sunday.

Write down a formula, in terms of v and w, for the money Vicky earned, £M, that week.

4   The diagram shows the plan of an L-shaped room. The dimensions of the room are given in metres.

Write down a formula, in terms of x and y, for the area, A square metres, of the room.

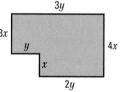

5   In the Russell family, there are 2 parents and 3 children. They are all going to London on the train. An adult train ticket from the Russell's local station to London costs £L and a child's ticket costs £l.

   a   Write down a formula, in terms of L and l, for the total cost (£T) of the family's tickets.

The next time the Russells plan to go to London, the train company has a special offer of 'a child travels half price if accompanied by an adult'.

   b   Write down a formula, in terms of L and l, for the total cost (£N) of their tickets this time.

19.7 Changing the subject of a formula

Exercise 19G

In each of the following formulae, change the subject to the letter given in brackets.

1 $a = 6b + 2$ (b)   **2** $x = 5y - 3$ (y)

3 $a = b - cd$ (d)   **4** $I = g + 4gh$ (h)

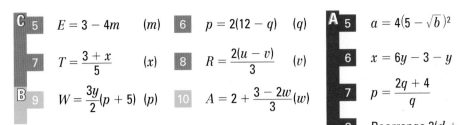

C 5 $E = 3 - 4m$ (m)

6   $p = 2(12 - q)$ (q)

7   $T = \dfrac{3 + x}{5}$ (x)

8   $R = \dfrac{2(u - v)}{3}$ (v)

B 9 $W = \dfrac{3y}{2}(p + 5)$ (p)

10   $A = 2 + \dfrac{3 - 2w}{3}$ (w)

19.8 Changing the subject in complex formulae

Exercise 19H

In each of the following formulae, change the subject to the letter given in brackets.

B 1 $V = \frac{1}{3}\pi r^2 h$ (r)

2   $P = \sqrt{x - y}$ (y)

3   $f = 3g^2 + 4h$ (g)

4   $T = \sqrt{\dfrac{2s}{g}}$ (s)

A 5 $a = 4(5 - \sqrt{b})^2$ (b)

6   $x = 6y - 3 - y$ (y)

7   $p = \dfrac{2q + 4}{q}$ (q)

8   Rearrange $3(d + 4) = c(1 - d)$ to make d the subject.

9   Make y the subject of $xy - 5z = 3 - 4y$.

10   Make N the subject of the formula $M = \left(\dfrac{7 - 4N}{N}\right)^2$.

20 Pythagoras' Theorem and trigonometry 1

🔍 Key Points

⊚ **Pythagoras' Theorem:** for a right-angled triangle the square of the hypotenuse equals the sum of the squares of the other two sides: $c^2 = a^2 + b^2$ or $AB^2 = BC^2 + CA^2$

⊚ **hypotenuse:** the longest side of a right angled triangle.

⊚ **adjacent side:** the side of a triangle next to a named angle.

⊚ **opposite side:** the side of a triangle opposite a named angle.

⊚ **sine (sin), cosine (cos) and tangent (tan):** the trigonometry ratios used to calculate angles and sides in right angled triangles.

$$\sin x° = \frac{opp}{hyp} \quad \cos x° = \frac{adj}{hyp} \quad \tan x° = \frac{opp}{adj}$$

⊚ **finding the hypotenuse:** add the square of each of the other sides and square root the answer. $c = \sqrt{(a^2 + b^2)}$.

⊚ **finding the length of a shorter side:** subtract the square of the other short side from the square of the hypotenuse, and square root the answer, e.g. $a = \sqrt{(c^2 - b^2)}$.

⊚ **finding the distance between two points on a coordinate grid:** the length of a line segment AB between A (x_1, y_1) and B (x_2, y_2) is $\sqrt{(x_2 - x_1)^2 + (y_2 - y_1)^2}$.

⊚ **using the trigonometry ratios to calculate lengths of sides:** used the rearranged trigonometry formulas
  ⊚ $opp = hyp \times \sin x°$
  ⊚ $adj = hyp \times \cos x°$
  ⊚ $opp = adj \times \tan x°$

20.1 Pythagoras' Theorem

⚙ Exercise 20A

Questions in this chapter are targeted at the grades indicated.

C **1** Work out the length of each hypotenuse marked with letters in these triangles.
Where appropriate, give each answer correct to 3 significant figures.

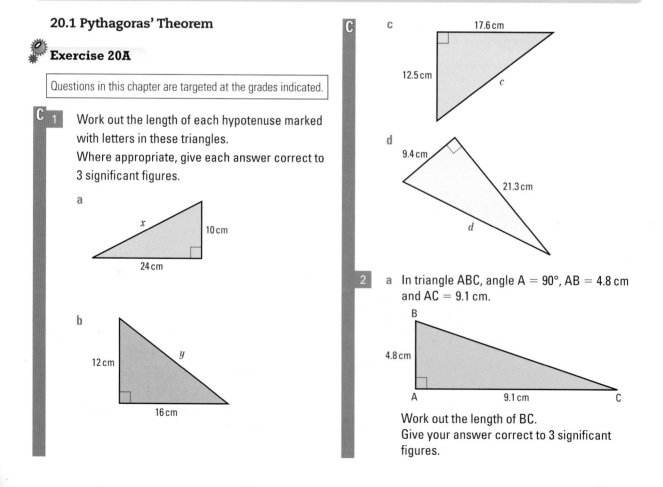

a

x, 10 cm, 24 cm

b

12 cm, y, 16 cm

C **c**

17.6 cm, 12.5 cm, c

d

9.4 cm, 21.3 cm, d

2 **a** In triangle ABC, angle A = 90°, AB = 4.8 cm and AC = 9.1 cm.

B, 4.8 cm, A, 9.1 cm, C

Work out the length of BC.
Give your answer correct to 3 significant figures.

C **b** In triangle XYZ, angle Y = 90°, XY = 7.4 cm and YZ = 7.9 cm.

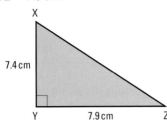

Work out the length of XZ.
Give your answer correct to 3 significant figures.

c In triangle GHI, angle G = 90°, GI = 11.8 cm and GH = 26.4 cm.

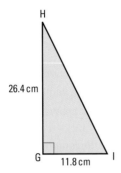

Work out the length of HI. Give your answer correct to 3 significant figures.

d In triangle STU, angle S = 90°, ST = 24.2 cm and SU = 30.1 cm.

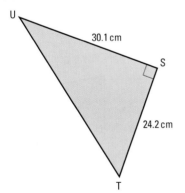

Work out the length of TU. Give your answer correct to 3 significant figures.

Exercise 20B

C **1** Work out the lengths of the sides marked with letters in these triangles.
Where appropriate give each answer correct to 3 significant figures.

ResultsPlus
Exam Tip

Check that the hypotenuse is the longest side of the triangle.

C **a**

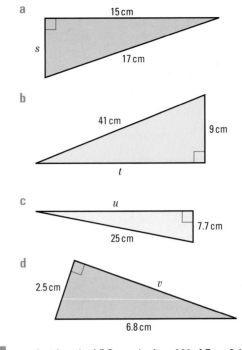

15 cm, 17 cm, s

b 41 cm, 9 cm, t

c u, 25 cm, 7.7 cm

d 2.5 cm, 6.8 cm, v

2 **a** In triangle ABC, angle A = 90°, AB = 8.1 cm and BC = 11.3 cm. Work out the length of AC. Give your answer correct to 3 significant figures.

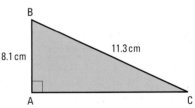

b In triangle XYZ, angle X = 90°, YZ = 22.4 cm and XZ = 19.2 cm.

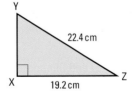

Work out the length of XY. Give your answer correct to 3 significant figures.

c In triangle PQR, angle R = 90°, PQ = 11.7 cm and PR = 3.8 cm.
i Draw a sketch of the right-angled triangle PQR and label sides PQ and PR with their lengths.
ii Work out the length of QR. Give your answer correct to 3 significant figures.

20.2 Applying Pythagoras' Theorem

Exercise 20C

C **1** Find the lengths of the sides marked with letters in each of these triangles.
Give each answer correct to 3 significant figures.

a

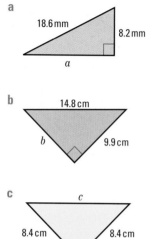

18.6 mm   8.2 mm a

b

14.8 cm b 9.9 cm

c

c 8.4 cm   8.4 cm

2 The diagram shows a rectangle of length 9 cm and width 6 cm.

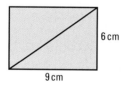

6 cm

9 cm

Work out the length of a diagonal of the rectangle. Give your answer correct to 3 significant figures.

3 Aiton (A), Beeville (B) and Ceaborough (C) are three towns as shown in this diagram.

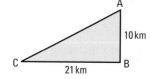

A

10 km

C   21 km   B

Beeville is 10 km due South of Aiton and 21 km due East of Ceaborough. Work out the distance between Aiton and Ceaborough.
Give your answer correct to the nearest km.

B **4**

R

2.5 m   5.5 m

P   S   M   2.3 m   T   Q

The diagram represents the decorated gable end of a roof. PSMTQ is a straight horizontal line. $\triangle$PQR and $\triangle$RST are both isosceles triangles.
SM = MT = 2.3 m, RM = 2.5 m and PR = RQ = 5.5 m.

a Work out the length of RS correct to 1 decimal place.

b Work out
  i MQ
  ii TQ.
Give your answers correct to 1 decimal place.

c What is the width of the roof (PQ)?
Give your answer correct to 1 decimal place.

5 Here are the lengths of sides of four triangles.

Triangle 1: 10 cm, 11 cm and 5 cm,
Triangle 2: 12 cm, 13 cm and 5 cm,
Triangle 3: 11 cm, 60 cm and 61 cm,
Triangle 4: 44 cm, 20 cm and 35 cm.

Which of these triangles are right-angled triangles?

6 The diagram shows two right-angled triangles.

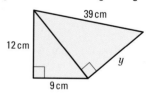

39 cm

12 cm

9 cm y

Work out the length of the side marked y.

7 A box, in the shape of a cuboid, has the following dimensions: length 24 cm, height 15 cm and width 10 cm. Each of the faces has a diagonal painted on it. What is the length of the diagonal, correct to the nearest cm, on:

a the top face
b an end face
c a side face?

20.3 Finding the length of a line segment

Exercise 20D

C 1
A02

Work out the length of the line joining each of these pairs of points. Where appropriate give your answer correct to 3 significant figures.

a (6, 2) and (11, 7)   b (5, 11) and (13, 30)
c (−3, 9) and (4, 13)   d (−3, −4) and (5, 11)
e (6, −17) and (−12, 7)   f (0, −6) and (9, −13)

2
A02
A03

The point X has coordinates (3, 2), the point Y has coordinates (6, 4) and the point Z has coordinates (1, 5).

a Work out the length of
   i XY   ii YZ   iii XZ
b What does your answer to part **a** tell you about triangle XYZ?

3
A02
A03

A circle has centre point O (2, 4).
The point P (6, 10) lies on the circle.
a Work out the radius of the circle.
b Determine by calculation which of the following points also lie on the circle.
   i S (14, 5)   ii T (−2, −9)
   iii U (5, 14)   iv V (8, 8)

20.4 Trigonometry in right-angled triangles

Exercise 20E

C 1

Use a calculator to find the value of

a sin 90°   b sin 62.6°
c cos 50°   d cos 28.7°
e tan 54°   f tan 68.3°
g cos 135.8°   h tan 5°
i sin 127.2°   j sin 41.7°
k tan 129.4°   l cos 78.8°

Give each answer correct to four decimal places, where necessary.

2

Use a calculator to find the value of x when

a $\cos x° = 0.7$ b $\sin x° = 0.5$
c $\cos x° = 0.41$ d $\tan x° = 0.65$
e $\sin x° = 0.5861$ f $\tan x° = 2.01$
g $\sin x° = 0.074$ h $\tan x° = \sqrt{2}$
i $\cos x° = \dfrac{\sqrt{3}}{2}$

Give each answer correct to 1 decimal place where necessary.

Exercise 20F

B 1

Write down which trigonometric ratio is needed to calculate either the length of the side marked p or the size of the angle marked x in each of these triangles. You do not have to calculate anything.

a

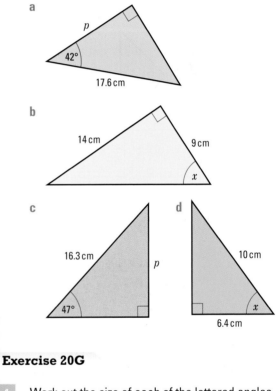

b

c   d

Exercise 20G

B 1

Work out the size of each of the lettered angles. Give each answer correct to one decimal place.

a

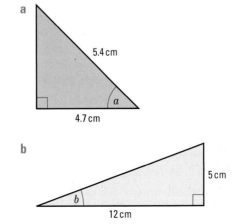

b

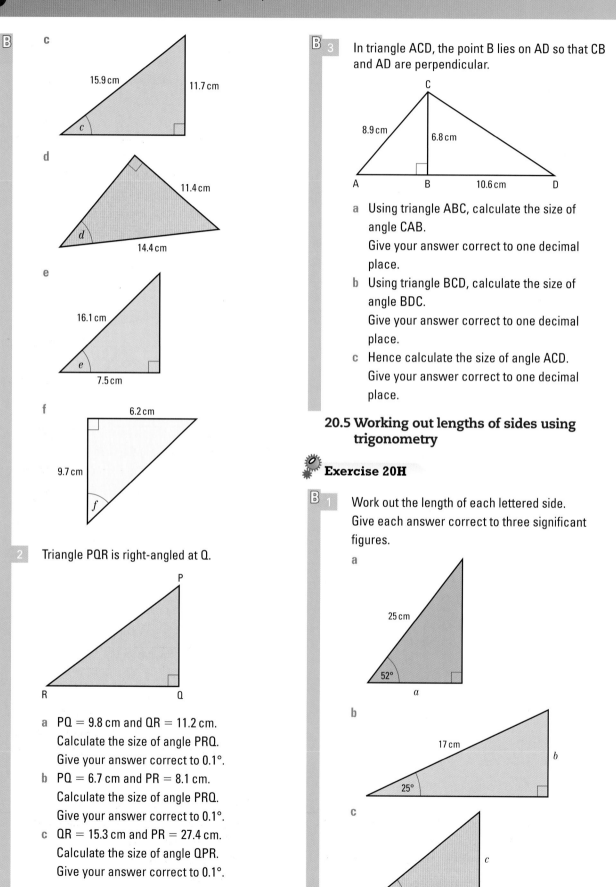

c

15.9 cm

11.7 cm

c

d

11.4 cm

d

14.4 cm

e

16.1 cm

e

7.5 cm

f

6.2 cm

9.7 cm

f

2 Triangle PQR is right-angled at Q.

P

R   Q

a PQ = 9.8 cm and QR = 11.2 cm.
Calculate the size of angle PRQ.
Give your answer correct to 0.1°.

b PQ = 6.7 cm and PR = 8.1 cm.
Calculate the size of angle PRQ.
Give your answer correct to 0.1°.

c QR = 15.3 cm and PR = 27.4 cm.
Calculate the size of angle QPR.
Give your answer correct to 0.1°.

3 In triangle ACD, the point B lies on AD so that CB and AD are perpendicular.

C

8.9 cm   6.8 cm

A   B   10.6 cm   D

a Using triangle ABC, calculate the size of angle CAB.
Give your answer correct to one decimal place.

b Using triangle BCD, calculate the size of angle BDC.
Give your answer correct to one decimal place.

c Hence calculate the size of angle ACD.
Give your answer correct to one decimal place.

20.5 Working out lengths of sides using trigonometry

Exercise 20H

1 Work out the length of each lettered side.
Give each answer correct to three significant figures.

a

25 cm

52°
a

b

17 cm

b

25°

c

c

39°

14.5 cm

B

d

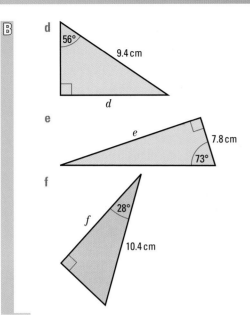

56°
9.4 cm
d

e

e
7.8 cm
73°

f

28°
f
10.4 cm

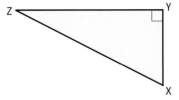

2 Triangle XYZ is right-angled at Y.

Z
Y
X

In each part, calculate the length of YZ.
Give each answer correct to three significant
figures.

a XY = 13.5 cm, angle YZX = 42°
b XZ = 16.7 cm, angle YXZ = 66°
c XZ = 11.4 m, angle YZX = 38°

3
A03 In triangle PQS, the point R lies on PS so that QR
and PS are perpendicular.

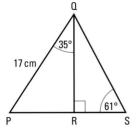

Q
35°
17 cm
61°
P   R   S

a Using triangle PQR, work out the length of
  i QR   ii PR.
  Give each answer correct to three significant
  figures.
b Using triangle QRS, work out the length of RS,
  correct to three significant figures.
c Hence calculate the length of PS, correct to
  three significant figures.
d Calculate the area of triangle PQS.
  Give your answer correct to the nearest cm².

Exercise 20I

B
1
A02 Where necessary give lengths correct to
3 significant figures and angles correct to
1 decimal place. A ladder is 5 m long. The ladder
rests against a vertical wall, with the foot of the
ladder resting on horizontal ground. The ladder
reaches up the wall a distance of 4.8 m.

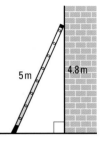

5 m   4.8 m

a Work out how far the foot of the ladder is from
  the bottom of the wall.
b Work out the angle that the ladder makes with
  the ground.

2
A02
A03 The diagram shows a vertical cliff face, PZ,
standing on a horizontal beach, XYZ.
The angle of elevation of P from a rock on the
beach at X is 44°, as shown in the diagram.

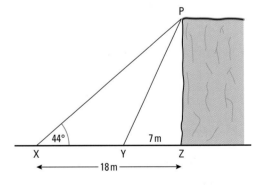

P
44°   7 m
X   Y   Z
18 m

a Work out the height, PZ, of the cliff face.
b Work out the angle of elevation of P from the
  point Y on the beach to the nearest degree.
c Work out the size of angle XPY to the nearest
  degree.

3 Mark and Holly are orienteering. Holly is 350 m
away from Mark. The bearing of Holly from Mark
is 125°. Work out how far:
a Holly is east of Mark
b Holly is south of Mark.

A 4

A02

The diagram shows the design for a logo that is to be hung outside a company's headquarters. It comprises triangles ABC and ADE.
Angle ACB = 90° = angle ADE.

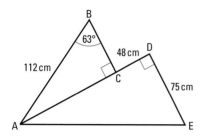

a What is the size of AC?
b Work out the size of angle DAE.

5

A02
A03

A rock, R, is 40 km from a harbour, H, on a bearing of 040°. A port, P, is 30 km from R on a bearing of 130°.
a Draw a sketch showing the points H, R and P and work out the size of angle HRP.
b Work out the distance HP.
c Work out the bearing of P from H.

A 6

A02
A03

The diagram shows an isosceles triangle.

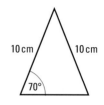

Calculate the area of the triangle.
Give your answer to the nearest cm².

21 More graphs and equations

Key Points

- **quadratic function:** an expression in which the highest power of x is x^2. They can be written in the form $ax^2 + bx + c$.
- **quadratic graph:** the graph of a quadratic function, known as a parabola. It has a smooth ⌢ or ⌣ shape with one line of symmetry.
- **minimum point:** the lowest point of a quadratic graph.
- **maximum point:** the highest point of a quadratic graph.
- **cubic function:** an expression in which the highest power of x is x^3. They can be written in the form $ax^3 + bx^2 + cx + d$.
- **cubic graph:** the graph of a cubic function, having one of the following shapes:

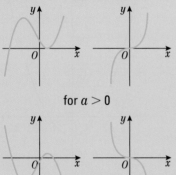

for $a > 0$

for $a < 0$

- **reciprocal function:** an expression of the form $\dfrac{k}{x}$.
- **reciprocal graph:** the graph of a reciprocal function. They are discontinuous and have two parts.

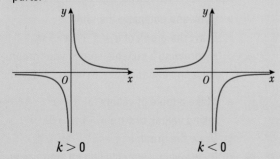

$k > 0$ $k < 0$

- **exponential functions:** expressions of the form a^x.
- **exponential graphs:** the graph of an exponential function. They are continuous, always lie above the x-axis, increase very quickly at one end and always cross the y-axis at $(0,1)$.

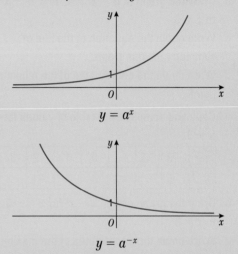

$y = a^x$

$y = a^{-x}$

- **trial and improvement:** a method used to find an approximate solution to an equation, if all other methods cannot be used.
- **solving quadratic equations of the form $ax^2 + bx + c = 0$:** read off the x-coordinate where the graph of $y = ax^2 + bx + c$ crosses the x-axis.
- **solving quadratic equations of the form $ax^2 + bx + c = mx + k$:** read off the x-coordinate at the point of intersection of the graph $y = ax^2 + bx + c$ with the straight line graph $y = mx + k$.
- **drawing cubic functions:** make a table of values then plot the points from your table and join them with a smooth curve.

21.1 Graphs of quadratic functions

Exercise 21A

Questions in this chapter are targeted at the grades indicated.

B 1 Here is the table of values for $y = x^2 - 2$.

x	-3	-2	-1	0	1	2	3
y	7		-1	-2			7

 a Copy and complete the table of values.
 b Draw the graph of $y = x^2 - 2$ for $x = -3$ to $x = 3$.
 c Write down the equation of the line of symmetry of your graph.
 d Write down the coordinates of the minimum point.

2 a Copy and complete the table of values for $y = 3 - x^2$.

x	-3	-2	-1	0	1	2	3
y	-6		2	3		-1	

 b Draw the graph of $y = 3 - x^2$ for $x = -3$ to $x = 3$.
 c Write down the coordinates of the maximum point.
 d Write down the values of x where the graph crosses the x-axis.

3 a Copy and complete the table of values for $y = 2x^2 + 1$.

x	-3	-2	-1	0	1	2	3
y		9		1	3	9	

 b Draw the graph of $y = 2x^2 + 1$ for $x = -3$ to $x = 3$.
 c Use your graph to find:
   i the value of y when $x = -2.5$
   ii the two values of x when $y = 6$.

4 Draw the graph for each of the following equations:
 a $y = x^2 + 3x - 2$ for values of x from -5 to 3
 b $y = 2x^2 - 5x + 3$ for values of x from -3 to 5
 c $y = (x - 3)^2$ for values of x from -1 to 7
 d $y = 6 + 2x - 2x^2$ for values of x from -3 to 5.
 For each case use your graphs to:
   i write down the value or values of x when the graph touches or crosses the x-axis
   ii draw in and write down the equation of the line of symmetry.

Exercise 21B

B 1 This is the graph of $y = 2x^2 - x - 6$.

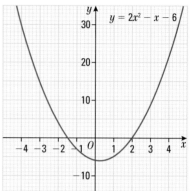

 Use the graph to solve the following equations.
 a $2x^2 - x - 6 = 0$
 A02 b $2x^2 - x - 3 = 0$

2 This is the graph of $y = 3x^2 - 5$.

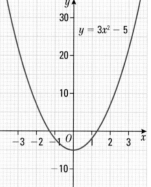

 Use the graph to solve the following equations.
 a $3x^2 - 5 = 0$
 A02 b $3x^2 - 2 = 0$

3 Here is a table of values for $y = 1 + 3x - x^2$.

x	-2	-1	0	1	2	3	4
y		-3	1	3			-3

 a Copy and complete the table.
 b Draw the graph of $y = 1 + 3x - x^2$.
 A03 c By drawing a suitable line on your graph, solve the equation $2 + 6x - 2x^2 = 2 - 2x$.

4 a Make a table of values for $y = x^2 - 4x + 4$, taking values of x from -1 to $+5$.
 b Draw the graph of $y = x^2 - 4x + 4$.
 A03 c By drawing a suitable line on your graph, solve the equation $x^2 - 2x = 0$.

21.2 Graphs of cubic functions

 Exercise 21C

1

a Copy and complete the table of values for $y = x^3 + 1$.

x	-3	-2	-1	0	1	2	3
y							

b Draw the graph of $y = x^3 + 1$ for $-3 \leqslant x \leqslant 3$.
c Use your graph to find the value of y when $x = 1.5$.

2 Here is a table of values for $y = x^3 - 5x$.

x	-4	-3	-2	-1	0	1	2	3	4
y		-12			0	-4		12	44

a Copy and complete the table.
b Draw the graph of $y = x^3 - 5x$ for $-4 \leqslant x \leqslant 4$.
c Use your graph to find the solutions to the equation $x^3 - 5x = 0$.

3

a Copy and complete the table of values for $y = 6x + x^2 - x^3$.

x	-3	-2	-1	0	1	2	3	4
y		0	-4			8	0	

b Draw the graph of $y = 6x + x^2 - x^3$ for $-3 \leqslant x \leqslant 4$.
c By drawing a suitable line on your diagram, solve the equation $6x + 2x^2 - x^3 = x - 2$.

4 Here are four graphs.
A

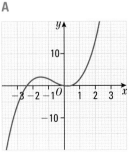

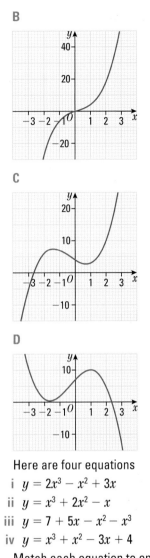

Here are four equations
i $y = 2x^3 - x^2 + 3x$
ii $y = x^3 + 2x^2 - x$
iii $y = 7 + 5x - x^2 - x^3$
iv $y = x^3 + x^2 - 3x + 4$
Match each equation to one of the graphs.
Give reasons for your answers.

21.3 Graphs of reciprocal functions

Exercise 21D

1

a Copy and complete the table of values for $y = \dfrac{4}{x}$ for $0 < x \leqslant 10$.

x	0.2	0.4	0.5	1	2	4	5	8	10
y		10		4	2		0.8		

b Using your answer to part **a**, copy and complete the following table of values for $y = \dfrac{4}{x}$ for $-10 \leqslant x < 0$.

x	-10	-8	-5	-4	-2	-1	-0.5	-0.4	-0.2
y									

c Draw the graph of $y = \dfrac{4}{x}$ for $-10 \leqslant x \leqslant 10$.

A **2** Draw the graph of $y = -\dfrac{3}{x}$ for $-10 \leqslant x \leqslant 10$.

A **3** A03
a Draw the graph of $y = \dfrac{8}{x+2}$ for $-3 \leqslant x \leqslant 8$.

b For which values of x is $y = \dfrac{8}{x+2}$ not defined?

21.4 Graphs of exponential functions

Exercise 21E

A **1**
a Copy and complete the table of values for $y = 4^x$. Give the values correct to 2 decimal places.

x	-3	-2	-1	0	1	2	3
y		0.06	0.25		4		64

b Draw the graph of $y = 4^x$ for $-3 \leqslant x \leqslant 3$.

c Use your graph to find an estimate for:
   i the value of y when $x = 1.5$
   ii the value of x when $y = 28$.

A **2** A03
The diagram shows the graphs of $y = 3^x$, $y = 2^{-x}$, $y = 5^x$ and $y = \left(\dfrac{1}{4}\right)^x$.

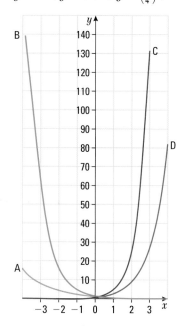

Match each graph to its equation.

A **3** A03
The number of rabbits, n, in a particular population grows at the rate given by the equation $n = 5 \times 2^y$ where y is the number of years.
a How many rabbits were there initially (when $y = 0$)?
b How many rabbits are there after 6 years?
c How many years will it take for the rabbit population to exceed 5000?

4 A03
The points (2, 100) and (0, 25) lie on the graph with equation $y = mn^x$ where m and n are constants. Work out the values of m and n.

21.5 Solving equations by the trial and improvement method

Exercise 21F

C **1** A02
For each of the following equations find two consecutive whole numbers between which a solution lies.

a $x^3 + x = 4$ b $x^3 + x^2 = 6$

c $x + \dfrac{2}{x^2} = 8$ d $x^2 - \dfrac{12}{x} = 0$

2
Use a trial and improvement method to find one solution of these equations correct to 1 decimal place.

a $x^3 + x = 5$ b $x^3 - x^2 + 6 = 0$

c $x + \dfrac{1}{x} = 4$

3
Use a trial and improvement method to find a positive solution of these equations correct to 2 decimal places.

a $x^3 - x = 32$ b $3x^2 - \dfrac{1}{x} = 13$

c $x^2(x + 2) = 140$

B **4** A02 A03
A box in the shape of a cuboid has a height of x cm. The length of the box is double the height. The width of the box is 2 cm more than the height. The volume of the box is 400 cm³.
a Show that x satisfies the equation $x^3 + 2x^2 = 200$.
b Use the trial and improvement method to solve the equation $x^3 + 2x^2 = 200$ correct to 1 decimal place.
c Write down the length, width and height of the box.

22 Quadratic and simultaneous equations

Key Points

- ⊚ **simultaneous equations:** two equations with two unknowns.
- ⊚ **prefect squares:** expressions such as $(x + 1)^2$, $(x + 4)^2$ and $(x + \frac{1}{2})^2$.
- ⊚ **completing the square:** a process for writing a quadratic expression so that the letter appears only in a squared term.
  - ⊚ expressions like $x^2 + bx + c$ can be written in the form $\left(x + \frac{b}{2}\right)^2 - \left(\frac{b}{2}\right)^2 + c$
  - ⊚ expressions like $ax^2 + bx + c$ are rewritten in the form $a\left(x^2 + \frac{b}{a}x\right) + c$ before completing the square on the expression in the brackets
- ⊚ **equation of a circle:** the equation of a circle of radius r and centre (0,0) is $x^2 + y^2 = r^2$
- ⊚ **solving simultaneous equations:**
  - ⊚ **elimination:** multiply the equations by a number so that the coefficients of an unknown are the same. Add or subtract the equations to eliminate the unknown. Solve the equation for the unknown that is left.
  - ⊚ **substitution:** rearrange one of the equations to make an unknown the subject. Substitute its value into the second equation so you have an equation in terms of the second unknown.
  - ⊚ **graphically:** draw the graphs of the two equations and find the coordinates of their point of intersection

- ⊚ **solving quadratic equations:** quadratic equations have two solutions that can be found using the following methods.
  - ⊚ **factorisation:** write the quadratic equation in the form $ax^2 + bx + c = 0$. Factorise the equation by finding two numbers whose sum is b and whose product is c. The solutions are found by writing each bracket = 0.
  - ⊚ **completing the square:** by completing the square any quadratic equation can be written in the form $p(x + q)^2 + r = 0$. Rearrange this equation to find the solutions.
  - ⊚ **using the formula:** all quadratic equations can be solved using the quadratic formula:
$$x = \frac{-b \pm \sqrt{b^2 - 4ac}}{2a}$$

22.1 Solving simultaneous equations

Exercise 22A

Questions in this chapter are targeted at the grades indicated.

Solve these simultaneous equations. Where appropriate, give your answers correct to two decimal places.

B
1. $x + y = 8$
    $x + 2y = 14$

2. $2x - y = 13$
    $6x + y = 3$

3. $3x - 2y = 9$
    $5x - 2y = 7$

4. $3x + 4y = 6$
    $x - 2y = 4$

B
5. $y = x + 6$
    $x + 3y = 9$

6. $2x + 4y = 12$
    $y = 2 - x$

7. $2x - y = -4$
    $y = 1 + 5x$

8. $4x - 3y = -2$
    $y = x + 1$

9. $y = 3 - 4x$
    $5x - 3y = -2$

A
10. $2x - 3y = 14$
    $4x + 2y = -7$

11. $2x - 3y = 1$
    $5x + 2y = 11$

12. $4x + 3y = 5$
    $6x + 4y = 4$

13. $3x + 4y = 5$
    $5x - 5y = -34$

14. $7x - 3y = 13$
    $4x - 2y = 13$

15. $2x + 3y = -1$
    $4x - 2y = 5$

22.2 Setting up equations in two unknowns

Exercise 22B

A 1 The sum of two numbers is 22 and their difference is 6.
Find the value of each of the numbers.

2 A02 A03 A garage bill was £346.50. The labour cost £187.60 more than the parts.
How much did the labour cost?

3 A03 Mr Brown bought 3 ice creams and 2 lollies. He paid £3.80. Jenny bought 2 ice creams and 4 lollies. She paid £4.40. If the ice creams cost x pence each and the lollies cost y pence each, work out the value of x and the value of y.

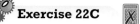

ResultsPlus
Watch Out!

Make sure that both sides of each equation have consistent units.

A★ 4 A02 A03 Four knives and six forks cost £13.60. Six knives and five forks cost £16.40.
Work out the cost of a knife and the cost of a fork.

5 A02 A03 Four adult tickets and three children's tickets for a children's zoo cost £92.60. Five adult tickets and four children's tickets cost £117.40.
How much would tickets for one adult and one child cost in total?

6 The diagram shows a rectangle.
All sides are measured in centimetres.

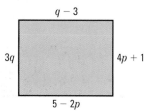

a Write down a pair of simultaneous equations in p and q.
b Solve your pair of simultaneous equations to find p and q.

A★ 7 A02 A03 A plumber charges a callout fee of £p plus £t for each 15 minutes he is working on a job. A job lasting 45 minutes costs £56.20. A job lasting $1\frac{1}{2}$ hours costs £98.50. What is the callout charge and the total cost of a job lasting $1\frac{1}{4}$ hours?

8 A02 A03 A train travels at 60 mph for x hours and 70 mph for y hours. It travels a distance of 290 miles and its average speed is 64 mph.
Find the times x and y to the nearest minute.

22.3 Using graphs to solve simultaneous equations

Exercise 22C

B 1 The diagram shows three lines **A**, **B** and **C**.

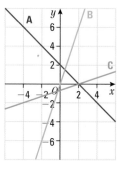

a Match the three lines to these equations.
$$y = 3x \qquad x + y = 2 \qquad x - 3y = 2$$
b Use the diagram to solve these simultaneous equations.
  i $y = 3x$
     $x + y = 2$
  ii $x + y = 2$
     $x - 3y = 2$
  iii $y = 3x$
      $x - 3y = 2$

2 For each of these pairs of simultaneous equations, draw two linear graphs on the same grid and use them to solve the equations.
Use a scale of -10 to $+10$ on each axis.
  a $x + y = 8$
     $4x - y = 2$
  b $3y = 2x - 3$
     $2x + y = 7$
  c $x + y = -4$
     $y = \frac{1}{2}x + 2$

22.4 Solving quadratic equations by factorisation

Exercise 22D

B **1** Solve

a $x(x - 5) = 0$ b $(y + 3)(y - 4) = 0$

c $(2z - 3)(4z - 7) = 0$ d $a^2 + 3a = 0$

e $t^2 - 2t = 0$ f $4g^2 - 5g = 0$

2 Solve

a $x^2 - 9x + 18 = 0$ b $x^2 + 10x + 16 = 0$

c $x^2 + x - 20 = 0$ d $x^2 - 10x + 24 = 0$

e $x^2 - 9x - 36 = 0$ f $x^2 - 25 = 0$

g $x^2 - 12x + 36 = 0$ h $x^2 - 81 = 0$

A **3** Solve

a $6x^2 + 37x + 6 = 0$ b $4x^2 - 19x + 12 = 0$

c $6x^2 + 17x - 3 = 0$ d $3x^2 + 11x - 4 = 0$

A* **4** Solve

a $x^2 - 2x = 8$ b $x^2 - 12 = 4x$

c $x(x - 1) = x + 15$ d $x^2 - 26 = 5x - 2$

e $3x(x - 1) = x^2 + 20$ f $3(2x^2 - 3) = 25x$

g $(x + 3)(x - 4) = 4(2x - 5)$

h $(2x - 1)^2 = 10 + x$

22.5 Completing the square

Exercise 22E

A **1** Write the following in the form $(x + p)^2 + q$.

a $x^2 + 8x$ b $x^2 + 14x$

c $x^2 + 20x$ d $x^2 - 4x$

e $x^2 - 16x$ f $x^2 - 22x$

g $x^2 + 2x$ h $x^2 - 5x$

i $x^2 + 8x + 15$ j $x^2 + 4x + 5$

k $x^2 + 10x - 18$ l $x^2 - 5x + 10$

m $x^2 - 20x + 60$ n $x^2 - 24x - 3$

o $x^2 - x + 3$ p $x^2 + 6x - 6$

2 Write the following in the form $a(x + p)^2 + q$.

a $2x^2 + 14x$ b $2x^2 - 6x + 5$

c $5x^2 + 40x + 80$ d $3x^2 - 9x + 10$

3 For all values of x, $x^2 + 6x + 18 = (x + p)^2 + q$.

a Find the value of the constants p and q.

b Write down the minimum value of
$x^2 + 6x + 18$.

A **4** The diagram shows a sketch of the curve with equation $y = x^2 + 4x + 12$.

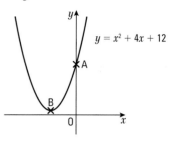

a Write down the coordinates of the point A, at which the curve crosses the y-axis.

b By completing the square for $x^2 + 4x + 12$, find the coordinates of the minimum point B.

A* **5** a By writing $2 + 2x - x^2$ as $-(x^2 - 2x - 2)$ find the value of p and the value of q for which
$2 + 2x - x^2 = p - (x - q)^2$.

b Using your answer to part **a**, what is the maximum value of $2 + 2x - x^2$?

c What is the value of x when this occurs?

22.6 Solving quadratic equations by completing the square

Exercise 22F

A **1** Solve these quadratic equations, giving your solutions in surd form.

a $x^2 + 4x + 2 = 0$ b $x^2 - 6x - 3 = 0$

c $x^2 + 8x - 10 = 0$ d $x^2 - 2x - 5 = 0$

e $3x^2 + 12x + 5 = 0$ f $2x^2 - 5 - 3 = 0$

2 Solve these quadratic equations, giving your solutions correct to 2 decimal places where appropriate.

a $x^2 + 7x + 5 = 0$ b $x^2 - 8x + 6 = 0$

c $x^2 + x - 6 = 0$ d $4x^2 + 2x - 5 = 0$

e $5x^2 - 10x - 4 = 0$ f $3x^2 - 6x - 2 = 0$

ResultsPlus
Exam Tip

Write down more digits than are required from your calculator display before you do any rounding.

22.7 Solving quadratic equations using the formula

Exercise 22G

Solve these quadratic equations. Where appropriate give your solutions correct to 3 significant figures.

A

1. $x^2 + 7x + 4 = 0$
2. $x^2 + 4x + 1 = 0$
3. $x^2 + 6x - 3 = 0$
4. $x^2 + x - 8 = 0$
5. $x^2 - 4x - 6 = 0$
6. $x^2 - 8x + 5 = 0$
7. $4x^2 + 6x + 1 = 0$
8. $3x^2 - 5x + 2 = 0$
9. $5x^2 - 9x - 1 = 0$
10. $2x^2 - 7x + 4 = 0$
11. $10x^2 + 2x - 3 = 0$
12. $x(x - 1) = x + 5$
13. $x(3x + 2) = 4 - 7x$
14. $(x + 2)(x - 3) = 12$
15. $7x - 3 = 4x^2 + 5(x - 2)$

22.8 Solving algebraic fraction equations leading to quadratic equations

Exercise 22H

Solve these quadratic equations.

A

1. $\dfrac{4}{x - 2} + \dfrac{6}{x + 2} = 1$
2. $\dfrac{8}{x} = \dfrac{4x - 10}{3}$
3. $\dfrac{5}{x} - \dfrac{6}{x + 1} = 2$
4. $\dfrac{6}{x} - \dfrac{4}{2x + 2} = 5$
5. $\dfrac{2}{3x - 1} + \dfrac{4}{x + 1} = 3$
6. $\dfrac{3}{2x - 3} + \dfrac{3}{x + 1} = 4$

In questions 7–12, give your solutions correct to 3 significant figures.

A

7. $\dfrac{4}{x} - \dfrac{1}{1 + x} = 1$
8. $\dfrac{3}{x + 4} - \dfrac{2}{x - 5} = 4$
9. $\dfrac{3}{x + 4} + \dfrac{5}{x - 1} = 1$
10. $\dfrac{2}{x - 1} - \dfrac{3}{x + 3} = 1$
11. $\dfrac{4x - 5}{x} = \dfrac{2 - x}{4}$
12. $\dfrac{6}{x - 1} + \dfrac{3}{1 - 5x} = 1$

22.9 Setting up and solving quadratic equations

Exercise 22I

A

1. The sum of the square of an integer and 3 times itself is 54.
   Find the two possible values of the integer.

2. The product of three unique numbers 6, $2x$ and $x - 8$ is -180.
   Find the value of the integer x.

3. A mother is five times as old as her daughter. Three years ago the product of their ages was 252. Find their ages this year as integers.

4. A garden's length is 6 m greater than its width. The area of the garden is 720 m^2.
   What is the length of the garden?

5. The sum of the squares of two consecutive integers is 85.
   a If x is one of the integers, show that $x^2 + x - 42 = 0$.
   b Solve $x^2 + x - 42 = 0$ to find the two consecutive integers.

6. Find the length of each side of this right-angled triangle.

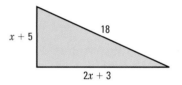

The measurements are given in cm.
Give your answers correct to 2 d.p.

7. A bathroom floor is being tiled. One size of square tile will cover the floor with 200 tiles. Another square tile, with sides 2 cm less, will use 250 tiles to cover the same area.
   Find the size of the larger tiles to the nearest cm.

8. A square painting canvas has sides of length 45 cm. A painter masks off a rectangular area on the canvas. She uses 90 cm of tape to mask three sides of the rectangle and the fourth side is formed by one edge of the canvas. The area enclosed in the rectangle is 1000 cm^2. Work out the length of the shorter sides of the rectangle.

9. A commemorative medal is made of a metallic disc set inside a ring made from a different metal. The diameter of the medal is $3x$ cm and the diameter of the inner disc is $(x + 1)$ cm. The area of the outer ring is $14\pi \text{ cm}^2$.
   What is the value of x?

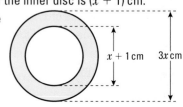

A* 10
A02
A03
A coach driver has to travel 560 km. The driver usually drives at an average speed of r km/h. He can arrive 45 minutes earlier if he increases his average speed by 6 km/h. Find the value of r correct to the nearest whole number.

22.10 Constructing graphs of simple loci

Exercise 22J

A 1
On graph paper, draw the graphs of the following equations.

a $x^2 + y^2 = 9$ b $x^2 + y^2 = 25$
c $x^2 + y^2 = 49$ d $x^2 + y^2 = 81$
e $x^2 + y^2 = 121$

2
A03
Using your graph of $x^2 + y^2 = 25$, construct a tangent parallel to the line $y = x$.
Write down the coordinates of the point where this tangent touches the graph.

A* 3
A03
Find the equation of the locus of points 4 units from the following lines.

a $y = 8$ b $y = -6$ c $x = 5$
d $x = -2$

4
A03
Find the equation of the locus of points 5 units from the line with equation $y = x - 3$.

22.11 Solving simultaneous equations when one is linear and the other is quadratic

Exercise 22K

For each of these pairs of simultaneous equations:

a draw a quadratic graph and a linear graph on the same grid and use them to solve the simultaneous equations (use a scale of -10 to $+10$ on each axis)

b solve them using an algebraic method. You must show all of your working.

Results Plus
Exam Tip

The graph of the linear equation crosses the graph of the quadratic at **two** points. So the pair of simultaneous equations has **two** solutions.

A 1
$x^2 + y = 24$
$y = 2x$

2
$x^2 - 2y = 15$
$y = 2x + 3$

3
$x^2 + 3y = 7$
$y + x = 1$

4
$y = 3x^2 - 2$
$y = 4 - 3x$

5
$y = 3 - x^2$
$y = 1 - x$

6
$2y = 4x^2 - 2$
$y = 3x + 1$

22.12 Solving simultaneous equations when one is linear and one is a circle

Exercise 22L

A* 1
On graph paper, draw the graph of the circle with equation $x^2 + y^2 = 16$. On the same axes, draw the straight line with equation $y = x$. Hence find estimates of the solutions of the simultaneous equations
$x^2 + y^2 = 16$ and $y = x$.

2
Draw suitable graphs to find estimates of the solutions of the simultaneous equations
$x^2 + y^2 = 27$ and $y = x + 2$.

3
Draw suitable graphs to find estimates of the solutions of the simultaneous equations
$x^2 + y^2 = 36$ and $x + y = 4$.

4
A02
Solve these simultaneous equations.

a $x^2 + y^2 = 20$ b $x^2 + y^2 = 34$
$\quad y = x + 2$ $\quad y = x + 8$
c $x^2 + y^2 = 4$
$\quad y = 2x + 4$

5
A02
Solve these simultaneous equations. Give your answers correct to 3 significant figures.

a $x^2 + y^2 = 30$ b $x^2 + y^2 = 24$
$\quad y = 3x + 1$ $\quad y = x + 3$
c $x^2 + y^2 = 100$
$\quad y = 2x - 5$

23 Area and volume 2

- **area of a circle sector:** if the circle has radius r and the sector has an angle of $x°$,
  area of a sector $= \frac{x}{360} \times \pi r^2$
- **volume:** the amount of space a 3D shape takes up.
  - volume of a pyramid $= \frac{1}{3} \times$ area of base $\times$ vertical height
  - volume of a cone $= \frac{1}{3} \times$ area of base $\times$ vertical height $= \frac{1}{3} \pi r^2 h$
  - volume of a sphere $= \frac{4}{3} \pi r^3$
- **surface area:** the area of the net that can be used to build the shape. Measured in square units.
  - surface area of a cylinder $= 2\pi rh + 2\pi r^2$
  - surface area of a cone $= \pi r^2 + \pi rl$
  - surface area of a sphere $= 4\pi r^2$

- **converting units of area:**

Length	Area
1 cm = 10 mm	$1\,cm^2 = 10 \times 10 = 100\,mm^2$
1 m = 100 cm	$1\,m^2 = 100 \times 100 = 10\,000\,cm^2$
1 km = 1000 m	$1\,km^2 = 1000 \times 1000 = 1\,000\,000\,m^2$

- **converting units of volume:**

Volume	
$1\,m^3 = 1\,000\,000\,cm^3$	$1\,cm^3 = \frac{1}{1\,000\,000}\,m^3$
$1\,cm^3 = 1000\,mm^3$	$1\,mm^3 = \frac{1}{1000}\,cm^3$
1 litre $= 1000\,cm^3$	$1\,cm^3 = 1\,ml$

- **working with coordinates in three dimensions:** three perpendicular axes are used, the x-axis, the y-axis and the z-axis. The coordinates of a point are written as (x, y, z).

23.1 Sectors of circles

Exercise 23A

Questions in this chapter are targeted at the grades indicated.

In this exercise, if your calculator does not have a π button, take the value of π to be 3.142.
Give answers correct to 3 significant figures unless the question says differently.

A **1** Calculate the arc length of each of these sectors.

a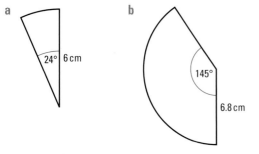

b

A **c**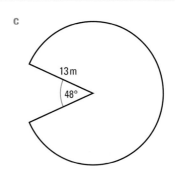

2 Calculate the area of each of these sectors.

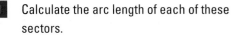

a

b

c

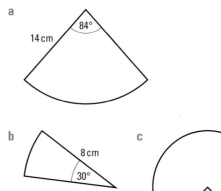

A **3** Calculate the perimeter of each of these sectors.

a

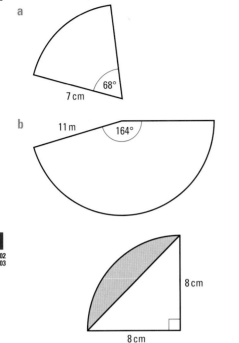

b

4

A02
A03

Calculate the area of the shaded segment.

A* **5**

A02
A03

The area of a sector of a circle is 75.4 cm².
The radius of the circle is 12 cm.
Calculate the length of the arc of the sector.

6

A02
A03

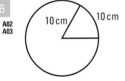

The radius of a circle is 10 cm. A sector of the circle has an arc length of 10 cm.
a   Work out the size of the angle of the sector.
b   Work out the area of the sector.

7

A02
A03

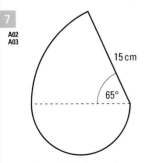

A semicircle and a sector of a circle make this shape. The sector has radius 15 cm and the angle of the sector is 65°. Work out:
a   the perimeter of the shape
b   the area of the shape.

23.2 Problems involving circles in terms of π

Exercise 23B

D **1** Giving your answer in terms of π, find the circumference of a circle:
a   with diameter 8 cm
b   with diameter 2 m
c   with radius 24 mm

2 Giving your answer in terms of π, find the area of a circle:
a   with radius 3 m
b   with radius 12 cm
c   with diameter 36 m

C **3**

A02
A03

This badge comprises two circles with the same centre.

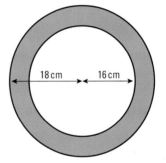

a   Show that the area of the coloured ring is 68π cm².
b   Find, in terms of π, the circumference of a circle whose area is the same as the coloured ring of the badge.

B **4** These two arcs make a complete circle.
The length of one arc is 8π cm and the length of the other arc is 12π cm.

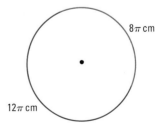

a   Find the radius of the circle.
b   Find, in terms of π, the area of the circle.

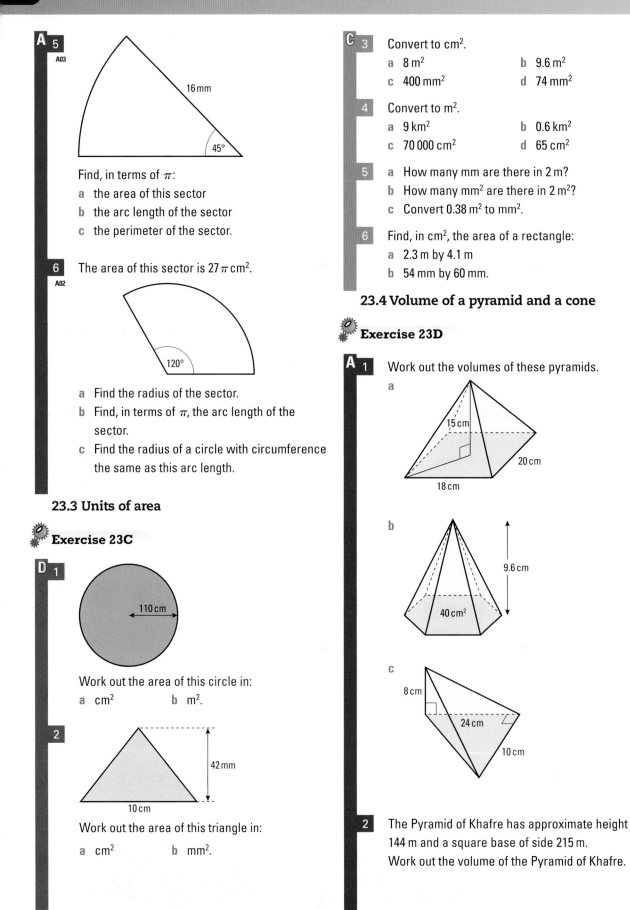

A 5

A03

Find, in terms of π:
a the area of this sector
b the arc length of the sector
c the perimeter of the sector.

6

A02

The area of this sector is $27\pi\,cm^2$.

a Find the radius of the sector.
b Find, in terms of π, the arc length of the sector.
c Find the radius of a circle with circumference the same as this arc length.

23.3 Units of area

Exercise 23C

D 1

Work out the area of this circle in:
a cm^2 b m^2.

2

Work out the area of this triangle in:
a cm^2 b mm^2.

C 3

Convert to cm^2.
a $8\,m^2$ b $9.6\,m^2$
c $400\,mm^2$ d $74\,mm^2$

4

Convert to m^2.
a $9\,km^2$ b $0.6\,km^2$
c $70\,000\,cm^2$ d $65\,cm^2$

5

a How many mm are there in 2 m?
b How many mm^2 are there in $2\,m^2$?
c Convert $0.38\,m^2$ to mm^2.

6

Find, in cm^2, the area of a rectangle:
a 2.3 m by 4.1 m
b 54 mm by 60 mm.

23.4 Volume of a pyramid and a cone

Exercise 23D

A 1

Work out the volumes of these pyramids.

a

b

c

2

The Pyramid of Khafre has approximate height 144 m and a square base of side 215 m.
Work out the volume of the Pyramid of Khafre.

3 Work out the volumes of these cones.
Use $\pi = 3.142$.
Give your answers correct to 3 significant figures.

a

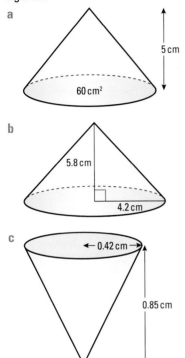

5 cm

60 cm²

b

5.8 cm

4.2 cm

c

← 0.42 cm →

0.85 cm

4 This is a child's toy made from two cones, one
with a base radius of 5 cm and a height of 5 cm
and the other with a base radius of 5 cm and a
height of 6 cm.

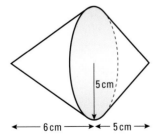

5 cm

← 6 cm → ← 5 cm →

Work out the total volume of the toy.
Leave your answer in terms of π.

5 The diagram shows a container that is made of a
cone cut into two parts.
Work out the volume of
both parts.

A

12 cm

← 14 cm →

B

← 20 cm →

6 cm

23.5 Volume of a sphere

Exercise 23E

In the following questions, use the value of π on your
calculator and give your answers correct
to 3 significant figures.

1 A spherical ball has a diameter of 45 cm.
Work out the volume of the ball.

2 Hayley makes a model of the northern
hemisphere of the Earth. Her model has a radius
of 15 cm. Work out the volume of the model.

3 The volume of a sphere is 3600 cm³.
Work out the radius of the sphere.

4 A sphere has the same volume as a cuboid box
with length 15 cm, width 30 cm and height 10 cm.
Work out the radius of the sphere.

5 A beach ball is made of 0.5 cm thick plastic.
The external diameter of the ball is 40.5 cm.
How many cubic centimetres of plastic are used
to make the ball?

6 The volume of a sphere of radius 3 metres is
twice the volume of a sphere of radius r metres.
Find the value of r.

23.6 Further volumes of shapes

Exercise 23F

1 Work out the volumes of these shapes. Give your
answers correct to 3 significant figures.

a

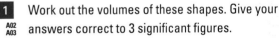

← 4 cm →

6 cm

4 cm

4 cm

4 cm

b

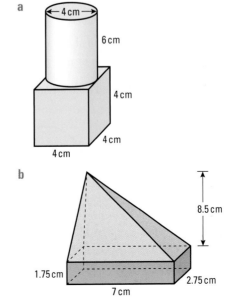

8.5 cm

1.75 cm

2.75 cm

7 cm

A

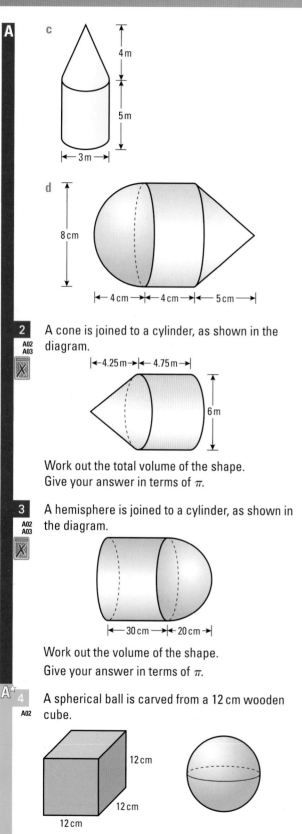

c

4 m

5 m

3 m

d

8 cm

4 cm — 4 cm — 5 cm

2 A cone is joined to a cylinder, as shown in the diagram.

4.25 m — 4.75 m

6 m

Work out the total volume of the shape.
Give your answer in terms of π.

3 A hemisphere is joined to a cylinder, as shown in the diagram.

30 cm — 20 cm

Work out the volume of the shape.
Give your answer in terms of π.

A **4** A spherical ball is carved from a 12 cm wooden cube.

12 cm

12 cm

12 cm

Work out the volume of wood that must be cut from the cube to make a ball with the largest possible radius.

A **5** A hemispherical hole is cut into a cylinder to make the shape shown in the diagram.

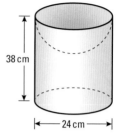

38 cm

24 cm

Work out the volume of the shape.
Give your answer in terms of π.

6 A shape is made by joining a hemisphere of radius $2r$ cm to a cone of radius $2r$ cm. The height of the cone is $5r$ cm.

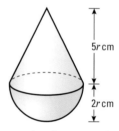

5r cm

2r cm

Find an expression, in terms of r and π, for the volume of the shape.

23.7 Units of volume

Exercise 23G

C **1** Convert these to cm³.
   a 4 m³   b 7.25 m³
   c 3600 mm³   d 8.6 mm³

2 Convert these to mm³.
   a 80 cm³   b 4.25 cm³
   c 0.07 cm³

3 Convert these to m³.
   a 2500 cm³   b 600 cm³
   c 175 000 mm³

4 Convert to litres.
   a 380 m*l*   b 45 000 cm³
   c 2.2 m³   d 5430 mm³

5 A water storage tank has a length of 4 m, a width of 8 m and a depth of 2.4 m.
How many litres of water does it hold?

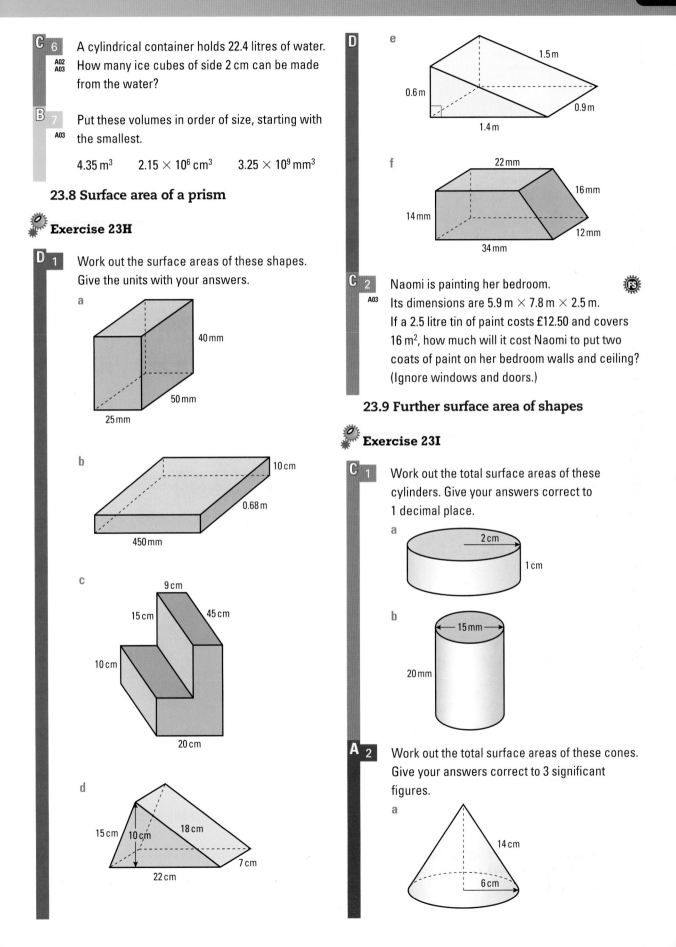

C 6
A02
A03
A cylindrical container holds 22.4 litres of water. How many ice cubes of side 2 cm can be made from the water?

B 7
A03
Put these volumes in order of size, starting with the smallest.

4.35 m³   2.15 × 10⁶ cm³   3.25 × 10⁹ mm³

23.8 Surface area of a prism

Exercise 23H

D 1
Work out the surface areas of these shapes. Give the units with your answers.

a

40 mm

50 mm

25 mm

b

10 cm

0.68 m

450 mm

c

9 cm

15 cm   45 cm

10 cm

20 cm

d

15 cm 10 cm

18 cm

7 cm

22 cm

D e

1.5 m

0.6 m

0.9 m

1.4 m

f

22 mm

16 mm

14 mm

12 mm

34 mm

C 2
A03
Naomi is painting her bedroom. Its dimensions are 5.9 m × 7.8 m × 2.5 m. If a 2.5 litre tin of paint costs £12.50 and covers 16 m², how much will it cost Naomi to put two coats of paint on her bedroom walls and ceiling? (Ignore windows and doors.)

23.9 Further surface area of shapes

Exercise 23I

C 1
Work out the total surface areas of these cylinders. Give your answers correct to 1 decimal place.

a

2 cm

1 cm

b

15 mm

20 mm

A 2
Work out the total surface areas of these cones. Give your answers correct to 3 significant figures.

a

14 cm

6 cm

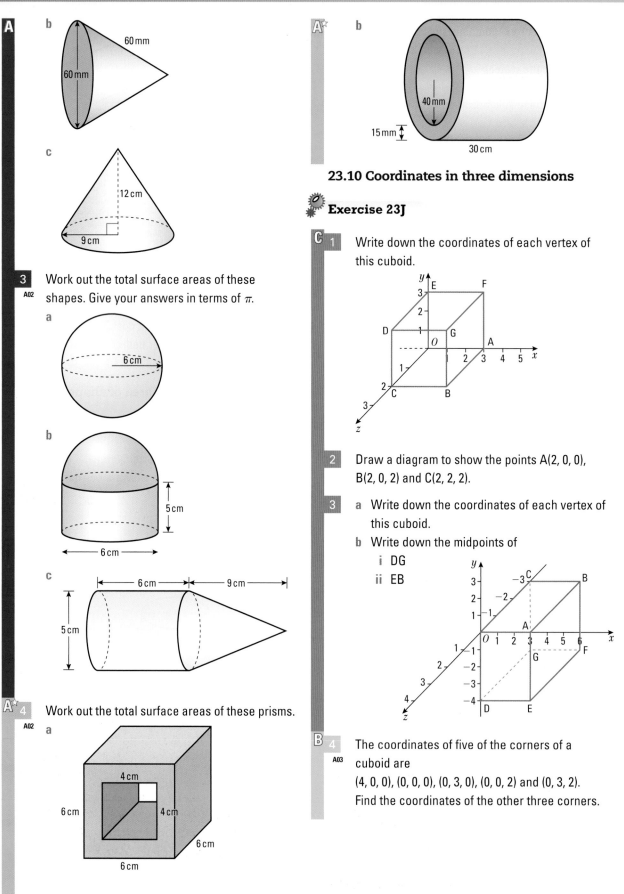

A

b

60 mm

60 mm

c

12 cm

9 cm

3 Work out the total surface areas of these shapes. Give your answers in terms of π.

A02

a

6 cm

b

5 cm

6 cm

c

6 cm   9 cm

5 cm

A **4** Work out the total surface areas of these prisms.

A02

a

4 cm

6 cm   4 cm

6 cm

6 cm

A **b**

40 mm

15 mm

30 cm

23.10 Coordinates in three dimensions

Exercise 23J

C **1** Write down the coordinates of each vertex of this cuboid.

2 Draw a diagram to show the points A(2, 0, 0), B(2, 0, 2) and C(2, 2, 2).

3 **a** Write down the coordinates of each vertex of this cuboid.

b Write down the midpoints of
  i DG
  ii EB

B **4** The coordinates of five of the corners of a cuboid are

A03

(4, 0, 0), (0, 0, 0), (0, 3, 0), (0, 0, 2) and (0, 3, 2).
Find the coordinates of the other three corners.

24 Line diagrams and scatter graphs

Key Points

- **bivariate data:** data that consists of pairs of related variables.
- **line graph:** a graph drawn from pairs of observations. Points are plotted on the graph and joined with straight lines.
- **scatter graph:** a graph of pairs of observations used to show whether there is any relationship between two variables.
- **correlation:** a relationship between pairs of variables.

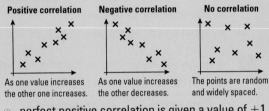

Positive correlation	Negative correlation	No correlation
As one value increases the other one increases.	As one value increases the other decreases.	The points are random and widely spaced.

- ⊙ perfect positive correlation is given a value of +1
- ⊙ perfect negative correlation is −1
- ⊙ if there is no correlation the correlation is 0

- **linear correlation:** when the points on a scatter graph lie approximately in a straight line.
- **line of best fit:** a straight line that passes as near as possible to the points on a scatter graph so as to best represent the trend of the graph.
  - ⊙ there should be approximately the same number of points on either side of the line
  - ⊙ further results can be estimated using the line of best fit
  - ⊙ **interpolation:** using the line of best fit to find values within the range of values on the scatter diagram
  - ⊙ **extrapolation:** using the line of best fit to find values outside the range of values on the scatter diagram
- **isolated point:** an extreme point that lies outside the normal range of values.

24.1 Drawing and using line graphs

Exercise 24A

Questions in this chapter are targeted at the grades indicated.

1 The line graph shows the time it took Zadie to drive to her mother's house.

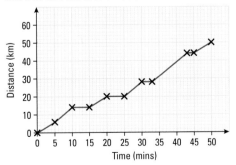

a How far was Zadie's journey to her mother's house?

b How many times did she stop on the journey and for how long did she stop?

c Use the line graph to estimate how long it took Zadie to drive 40 km.

d Use the line graph to estimate how far she had driven in 35 minutes.

2 The line graph shows the depth, in cm, of snow on the upper slopes of a ski resort each month for one year.

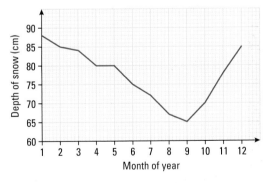

a When was the depth of snow 85 cm?

b When was the deepest snow? How deep was the snow in this month?

c Use the line graph to estimate the depth of snow in the 11th month.

A03 d What was the lowest depth of snow? In which month did this occur? Suggest a reason for this.

3 The line graph shows the water used in a family household over one week.

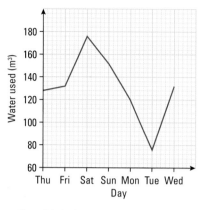

a On which day was consumption highest? Suggest a reason for this.
b How much water was used on Sunday?
c Use the line graph to estimate the days when 132 m³ of water was used.
d When was least water used? Suggest a reason for this.

24.3 Recognising correlation

Exercise 24B

1 This table gives information about the depth of tyre treads, in millimetres, and the stopping distance, in metres, of ten cars.

Tread depth (mm)	Stopping distance (m)
4	30
5	32
2.5	32
6.5	28
3.5	34
1	37
1	49
1.5	40
1.2	42
2.5	35

a Draw a scatter graph for the given data.
b Describe the correlation.
c Describe the relationship between the depth of tyre treads and the stopping distance of a car.

2 This table gives information about the selling price of microwave ovens and their capacity in litres.

Capacity (litres)	Selling price (£)
20	45
17	27
23	68
25	73
20	54
17	34
18	30
24	70

a Draw a scatter graph for the given data.
b Describe the correlation.
c What do you notice about capacity and cost?

3 A survey has been done on the affect of birth rate on life expectancy in 10 different countries.

Birth rate per 1000 population	Life expectancy (years)
10	70
14	74
19	72
20	68
35	65
50	42
30	60
40	55
48	48
43	50

a Draw a scatter graph for the given data.
b Describe the correlation.
c Describe the relationship between birth rate and life expectancy in these countries.

4 Describe the relationship between the following pairs of variables, giving your reasons.
a The number of cars sold and the number of satellite navigation systems bought.
b The number of dishwashers and the number of televisions sold in a department store.
c The length of a lorry driver's journey and the number of stops he makes.
d The weight of a woman and the hours she spends at the gym.

24.4 Drawing lines of best fit

Exercise 24C

D **1** The scatter graph shows the best and worst prices offered on the Internet for 12 used mobile phones.

a Copy the scatter graph.
b One point seems to be an isolated point. Circle the isolated point and write down its coordinates.
c Ignoring the isolated point, draw a line of best fit on your scatter graph.

2 The table shows the ages and prices of 10 second-hand cars.

Age (years)	Price (£1000)
1	8
2	7
3	7.5
4	6
5	4
2	9
6	3
7	2.5
8	1.5
9	0.5

a Draw a scatter graph for these data.
b One point seems to be an isolated point. Circle it and suggest a reason why this might have occurred.
c Ignoring the isolated point, draw a line of best fit on your scatter graph.

D **3** The table shows information that has been collected by a researcher about the highest and the lowest temperatures recorded in a 24-hour period in some cities. The first six cities have been plotted on the scatter graph.

	High temp. (°F)	Low temp. (°F)
Athens	72	59
Barcelona	51	40
Brussels	43	31
Bucharest	53	32
Cairo	68	53
Dubai	74	64

	High temp. (°F)	Low temp. (°F)
Florence	47	39
Lisbon	59	43
Marseilles	61	54
Prague	39	45
Sofia	45	37
Zurich	40	30

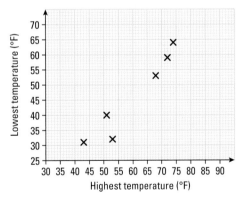

a Copy the scatter graph and complete it by plotting the last six points.
b There is one point that seems to be an isolated point. Circle this point.
c Write down the name of the city that is the isolated point. Suggest a reason for this isolated point.
d Ignoring the isolated point, draw a line of best fit on your scatter graph.

24.5 Using lines of best fit to make predictions

Exercise 24D

C 1
A03
The scatter graph shows the marks in physics and chemistry of a group of students.

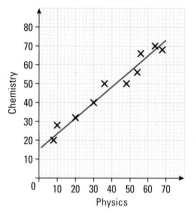

a A student gets a mark of 40 in physics. Use the line of best fit to find the mark she is likely to get in chemistry.

b A student gets a mark of 68 in chemistry. Use the line of best fit to find the mark he is likely to get in physics.

2
A03
The scatter graph shows temperature at different altitudes in a certain country.

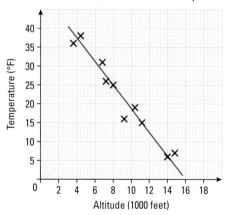

a Use the line of best fit to work out an estimate of the temperature at an altitude of:
　i 12 000 feet　　ii 6000 feet.

b Use the line of best fit to work out an estimate of the altitude where you are likely to get a temperature of:
　i 10°F　　　　　ii 35°F.

C 3
A02
A03
The high jumps of eight female athletes were recorded against their height at an athletics meeting.

Height of athlete (cm)	Height of jump (m)
155	2.01
170	2.07
160	2.06
180	2.09
160	2.05
175	2.07
153	2.03
150	2.00

a Draw a scatter graph for the given data.

b Work out an estimate of the height jumped when an athlete's height is
　i 173 cm　　ii 165 cm.

c Work out an estimate of the height of an athlete jumping 2.04 m.

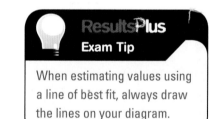

Results**Plus**
Exam Tip

When estimating values using a line of best fit, always draw the lines on your diagram.

25 Indices, standard form and surds

Key Points

⦿ **standard form:** a number written in the form $a \times 10^n$ where $1 \leq a < 10$ and n is an integer.

   ◉ used to represent very large or very small numbers

   ◉ use the $\boxed{10^x}$ or $\boxed{\text{EXP}}$ key on your calculator

⦿ **fractional index rules:**

   ◉ $a^0 = 1$

   ◉ $a^{-n} = \dfrac{1}{a^n}$

   ◉ $a^{\frac{1}{n}} = \sqrt[n]{a}$

⦿ **surd:** a number written exactly using square roots, e.g. $\sqrt{2}$ and $\sqrt{3}$.

⦿ **simplifying surds:** use these two rules:

   ◉ $\sqrt{m} \times \sqrt{n} = \sqrt{mn}$

   ◉ $\dfrac{\sqrt{m}}{\sqrt{n}} = \sqrt{\dfrac{m}{n}}$

⦿ **rationalising the denominator:** means to get rid of any surds in the denominator. To rationalise the denominator of $\dfrac{a}{\sqrt{b}}$ multiply the fraction by $\dfrac{\sqrt{b}}{\sqrt{b}}$.

25.1 Using zero and negative powers

Exercise 25A

Questions in this chapter are targeted at the grades indicated.

1 Write down the value of these expressions.

  a 9^0 b 6^{-1} c 4^{-1} d 5^0

  e $(-3)^{-3}$ f 7^{-2} g 10^{-5} h 182^0

  i $(-2)^{-2}$ j $(-12)^0$ k 18^0 l 10^{-8}

2 Work out the value of these expressions.

  a $\left(\dfrac{2}{3}\right)^{-1}$ b $\left(\dfrac{2}{5}\right)^{-1}$ c $\left(\dfrac{2}{7}\right)^{-2}$ d $\left(\dfrac{3}{4}\right)^{-3}$

  e $(0.75)^{-2}$ f $\left(\dfrac{3}{10}\right)^{-3}$ g $\left(\dfrac{4}{3}\right)^{0}$ h $\left(\dfrac{7}{5}\right)^{-1}$

  i $\left(1\dfrac{3}{5}\right)^{-2}$ j $\left(1\dfrac{2}{3}\right)^{-3}$ k $(0.4)^{-3}$ l $(0.2)^{-4}$

25.2 Using standard form

Exercise 25B

1 Write these numbers in standard form.

  a 60 000   b 3000   c 700

  d 400 000   e 800 000 000

2 Write these as ordinary numbers.

  a 2×10^4 b 8×10^8 c 1×10^5

  d 6×10^1 e 4×10^5

3 Write these numbers in standard form.

  a 652 000   b 50   c 45

  d 34 000   e 43.8

4 Write these as ordinary numbers.

  a 7.02×10^3 b 8.6×10^7 c 6.94×10^4

  d 7.8×10^1 e 3.29×10^0

5 In 2008 there were approximately 61 000 000 people in the UK.
Write this number in standard form.

6 The diameter of Earth is approximately 13 000 km.
Write this number in standard form.

Exercise 25C

1 Write these numbers in standard form.

  a 0.03   b 0.0007   c 0.006

  d 0.000 005   e 0.8

2 Write these as ordinary numbers.

  a 3×10^{-7} b 6×10^{-5} c 2×10^{-8}

  d 5×10^{-2} e 1×10^{-1}

3 Write these numbers in standard form.

  a 0.106   b 0.000 61   c 0.0074

  d 0.678   e 0.000 408 2

4 Write these as ordinary numbers.

  a 1.03×10^{-2} b 3.84×10^{-5} c 3.4×10^{-7}

  d 4.9×10^{-4} e 4.582×10^{-1}

5 Write these numbers in standard form.

  a 98 000   b 0.0035   c 0.287

  d 300   e 0.002 64   f 745 000

  g 0.000 007 48   h 4063   i 0.0004

  j 3 562 000

B 6 Write these as ordinary numbers.
a 1.25×10^{-7} b 7.2×10^5 c 1.44×10^{-4}
d 4×10^3 e 5.7×10^{-1} f 2.73×10^8
g 4.3215×10^2 h 5×10^{-6} i 2.61×10^{-1}
j 3.052×10^7

7 1 micron is 0.000 001 of a metre. Write down the size of 3 microns, in metres, in standard form.

8 A red blood cell has a diameter of 0.0076 mm. Write this number in standard form.

Exercise 25D

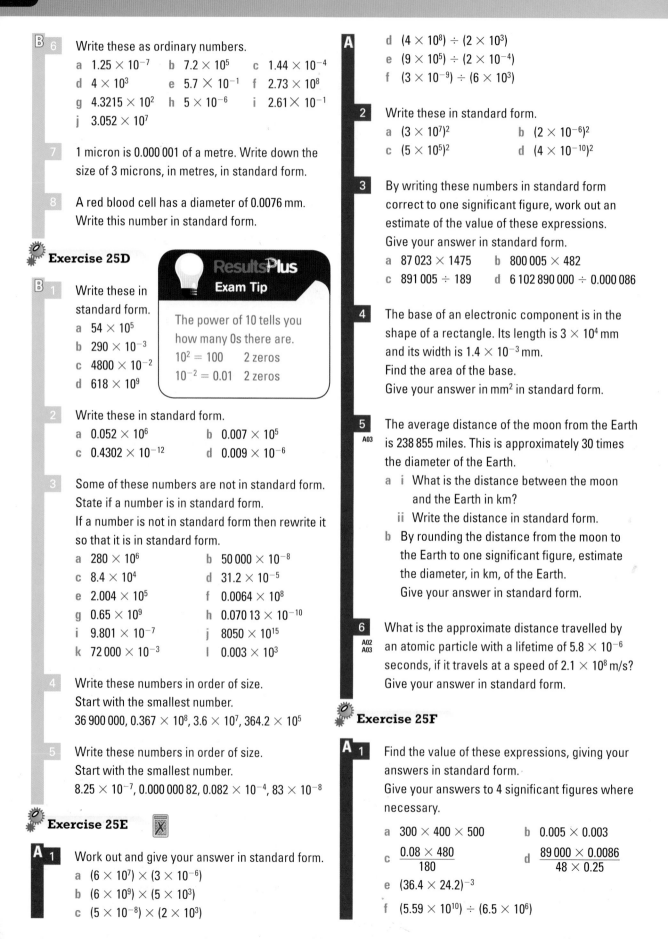

ResultsPlus
Exam Tip

The power of 10 tells you how many 0s there are.
$10^2 = 100$ 2 zeros
$10^{-2} = 0.01$ 2 zeros

B 1 Write these in standard form.
a 54×10^5
b 290×10^{-3}
c 4800×10^{-2}
d 618×10^9

2 Write these in standard form.
a 0.052×10^6 b 0.007×10^5
c 0.4302×10^{-12} d 0.009×10^{-6}

3 Some of these numbers are not in standard form. State if a number is in standard form.
If a number is not in standard form then rewrite it so that it is in standard form.
a 280×10^6 b $50\,000 \times 10^{-8}$
c 8.4×10^4 d 31.2×10^{-5}
e 2.004×10^5 f 0.0064×10^8
g 0.65×10^9 h $0.070\,13 \times 10^{-10}$
i 9.801×10^{-7} j 8050×10^{15}
k $72\,000 \times 10^{-3}$ l 0.003×10^3

4 Write these numbers in order of size. Start with the smallest number.
$36\,900\,000$, 0.367×10^8, 3.6×10^7, 364.2×10^5

5 Write these numbers in order of size. Start with the smallest number.
8.25×10^{-7}, $0.000\,000\,82$, 0.082×10^{-4}, 83×10^{-8}

Exercise 25E

A 1 Work out and give your answer in standard form.
a $(6 \times 10^7) \times (3 \times 10^{-6})$
b $(6 \times 10^9) \times (5 \times 10^3)$
c $(5 \times 10^{-8}) \times (2 \times 10^3)$

A
d $(4 \times 10^8) \div (2 \times 10^3)$
e $(9 \times 10^5) \div (2 \times 10^{-4})$
f $(3 \times 10^{-9}) \div (6 \times 10^3)$

2 Write these in standard form.
a $(3 \times 10^7)^2$ b $(2 \times 10^{-6})^2$
c $(5 \times 10^5)^2$ d $(4 \times 10^{-10})^2$

3 By writing these numbers in standard form correct to one significant figure, work out an estimate of the value of these expressions. Give your answer in standard form.
a $87\,023 \times 1475$ b $800\,005 \times 482$
c $891\,005 \div 189$ d $6\,102\,890\,000 \div 0.000\,086$

4 The base of an electronic component is in the shape of a rectangle. Its length is 3×10^4 mm and its width is 1.4×10^{-3} mm.
Find the area of the base.
Give your answer in mm² in standard form.

5 [A03] The average distance of the moon from the Earth is 238 855 miles. This is approximately 30 times the diameter of the Earth.
a i What is the distance between the moon and the Earth in km?
  ii Write the distance in standard form.
b By rounding the distance from the moon to the Earth to one significant figure, estimate the diameter, in km, of the Earth.
Give your answer in standard form.

6 [A02 A03] What is the approximate distance travelled by an atomic particle with a lifetime of 5.8×10^{-6} seconds, if it travels at a speed of 2.1×10^8 m/s?
Give your answer in standard form.

Exercise 25F

A 1 Find the value of these expressions, giving your answers in standard form.
Give your answers to 4 significant figures where necessary.
a $300 \times 400 \times 500$ b 0.005×0.003
c $\dfrac{0.08 \times 480}{180}$ d $\dfrac{89\,000 \times 0.0086}{48 \times 0.25}$
e $(36.4 \times 24.2)^{-3}$
f $(5.59 \times 10^{10}) \div (6.5 \times 10^6)$

A

g $(1.3 \times 10^7) \div (4.5 \times 10^6)$

h $(1.52 \times 10^{10}) \div (4.75 \times 10^6)$

i $(2.4 \times 10^{-8}) \div (3.5 \times 10^{-6})$

2 Evaluate these expressions. Give your answers in standard form correct to 3 significant figures.

a $(3.24 \times 10^8) \div (6.4 \times 10^4)$

b $(5.3 \times 10^8) \times (6.45 \times 10^6)$

c $(3.24 \times 10^{-8}) \div (6.4 \times 10^{-4})$

d $(5.3 \times 10^{-8}) \times (6.45 \times 10^{-6})$

3 $x = 5.3 \times 10^9$, $y = 6.4 \times 10^4$

Work out the following. Give your answer in standard form correct to 3 significant figures.

a $\dfrac{x}{y}$ b $x(2x + 300y)$

c $\dfrac{xy}{400x + y}$ d $\left(\dfrac{x}{1000}\right)^2 + y^2$

4 $x = 4.3 \times 10^{-4}$, $y = 6.9 \times 10^{-6}$

Evaluate these expressions.

Give your answer in standard form correct to 3 significant figures where necessary.

a $\dfrac{x^2}{2y}$ b $\dfrac{x^2 + y^2}{x + y}$ c $\dfrac{x + y}{xy}$

5 The distance of the Earth from the Sun is 1.5×10^8 km.

The distance of the planet Saturn from the Sun is 1430 million km.

Write in the form $1 : n$ the ratio

distance of the Earth from the Sun : distance of the planet Saturn from the Sun

6 The mass of a potassium atom is 6.49×10^{-23} grams.

Work out the number of potassium atoms in 2.75 kilograms of potassium.

25.3 Working with fractional indices

Exercise 25G

B 1 Work out the value of the following.

a $16^{\frac{1}{2}}$ b $64^{\frac{1}{2}}$ c $225^{\frac{1}{2}}$

d $25^{\frac{1}{2}}$ e $\left(\frac{1}{9}\right)^{\frac{1}{2}}$

2 Work out the value of

a $8^{\frac{1}{3}}$ b $64^{\frac{1}{3}}$ c $(-125)^{\frac{1}{3}}$

d $(-1000)^{\frac{1}{3}}$ e $\left(\frac{1}{27}\right)^{\frac{1}{3}}$

B 3 Work out the value of

a $81^{-\frac{1}{4}}$ b $36^{-\frac{1}{2}}$ c $8^{-\frac{1}{3}}$

d $\left(\frac{1}{100\,000}\right)^{-\frac{1}{5}}$ e $\left(\frac{16}{25}\right)^{-\frac{1}{2}}$

A 4 Work out the value of

a $8^{\frac{2}{3}}$ b $64^{\frac{2}{3}}$ c $125^{\frac{2}{3}}$

d $10\,000^{\frac{3}{4}}$ e $16^{\frac{3}{2}}$

5 Work out, as a single fraction, the value of

a $64^{-\frac{2}{3}}$ b $16^{-\frac{3}{4}}$ c $1000^{-\frac{1}{3}}$

d $27^{-\frac{2}{3}}$ e $81^{-\frac{3}{2}}$

f $64^{-\frac{2}{3}} \times \left(\frac{1}{4}\right)^2$ g $27^{-\frac{1}{3}} \times \left(\frac{3}{5}\right)^2$

A* 6 Find the value of n.

a $32 = 2^n$ b $\frac{1}{6} = 6^n$ c $(\sqrt{5})^7 = 5^n$

d $\dfrac{1}{\sqrt[3]{8}} = 8^n$ e $(\sqrt[3]{3})^{10} = 3^n$

25.4 Using surds

Exercise 25H

A 1 Find the value of the integer k.

a $\sqrt{12} = k\sqrt{3}$ b $\sqrt{27} = k\sqrt{3}$

c $\sqrt{75} = k\sqrt{3}$ d $\sqrt{48} = k\sqrt{3}$

2 Simplify

a $\sqrt{48}$ b $\sqrt{300}$ c $\sqrt{24}$ d $\sqrt{50}$

3 Solve the equation $x^2 = 40$, leaving your answer in surd form.

A* 4 Expand these expressions.

Write your answers in the form $a + b\sqrt{c}$ where a, b and c are integers.

a $\sqrt{5}(2 + \sqrt{5})$ b $(\sqrt{5} + 4)(1 + \sqrt{5})$

c $(\sqrt{3} + 1)(2 - \sqrt{3})$ d $(\sqrt{11} - 2)(3 + \sqrt{11})$

e $(3 - \sqrt{5})^2$ f $(4 + \sqrt{6})^2$

5 The area of a square is 2420 cm².

Find the length of one side of the square.

Give your answer as a surd in its simplest form.

6 The lengths of the sides of a rectangle are $(7 - \sqrt{2})$ cm and $(\sqrt{2} + 7)$ cm.

Work out, in their simplified forms:

a the perimeter of the rectangle

b the area of the rectangle.

A★ 7 The length of the side of a square is $(4 + \sqrt{3})$ cm.
Work out the area of the square.
Give your answer in the form $(a + b\sqrt{3})$ cm²
where a and b are integers.

 Exercise 25I

A 1 Rationalise the denominators and simplify your
answers, if possible.

a $\dfrac{2}{\sqrt{7}}$ b $\dfrac{5}{\sqrt{3}}$ c $\dfrac{5}{\sqrt{11}}$

d $\dfrac{10}{\sqrt{2}}$ e $\dfrac{15}{\sqrt{3}}$

A★ 2 Rationalise the denominators and give your
answers in the form $a + b\sqrt{c}$ where a, b and c
are integers.

a $\dfrac{6 + \sqrt{3}}{\sqrt{3}}$ b $\dfrac{10 - \sqrt{5}}{\sqrt{5}}$ c $\dfrac{10 + \sqrt{2}}{\sqrt{2}}$

d $\dfrac{15 + \sqrt{3}}{\sqrt{3}}$ e $\dfrac{12 - \sqrt{2}}{\sqrt{2}}$

A★ 3 The diagram shows a right-angled triangle.
The lengths are given in centimetres.

The diagram shows a right-angled triangle with vertical side $\dfrac{2}{\sqrt{2}}$ and horizontal side $\dfrac{12}{\sqrt{3}}$.

Work out the area of the triangle.
Give your answer in the form $a + b\sqrt{c}$ where a, b
and c are integers.

4 Solve these equations leaving your answers in
surd form.

a $x^2 - 6x + 4 = 0$ b $x^2 + 12x + 8 = 0$

5 The diagram represents a right-angled triangle
XYZ.
$XY = (\sqrt{5} - 2)$ cm $YZ = (\sqrt{5} + 2)$ cm.

Triangle XYZ with the right angle at Y, $XY = (\sqrt{5} - 2)$ and $ZY = (\sqrt{5} + 2)$.

Work out, leaving any appropriate answers in
surd form:

a the area of triangle XYZ

b the length of XZ.

26 Similar shapes

Key Points

⊙ **linear scale factor (k):** the ratio of the corresponding sides in two similar shapes.
  ⊙ **area scale factor:** k^2
  ⊙ **volume scale factor:** k^3

⊙ **finding the length, area and volume of similar shapes:**
  ⊙ length in similar shape = corresponding length in original shape × scale factor
  ⊙ area of similar shape = area of original shape × scale factor²
  ⊙ volume of similar shape = volume of original shape × scale factor³

26.1 Areas of similar shapes

Exercise 26A

Questions in this chapter are targeted at the grades indicated.

A **1** Shapes A and B are similar.
Shape A has area 12.25 cm² and length 3.5 cm.
Shape B has length 17.5 cm.

A
← 3.5 cm →

B
← 17.5 cm →

Calculate the area of shape B.

2 **A02** Triangles P and Q are similar.
The area of triangle Q is 192 cm².
Calculate the area of triangle P.

P
← 2.5 cm →

Q
← 20 cm →

3 **A02** Cylinders R and S are similar.
The surface area of cylinder R is 90 cm².

3.2 cm R

11.2 cm S

Calculate the surface area of cylinder S.

A **4** **A02** A birthday cake is made up of three similar square cakes. Each cake has a ribbon tied round it. The smallest cake has a surface area of 726 cm² and its ribbon is 60 cm long. The other two cakes have ribbons of lengths 180 cm and 200 cm.

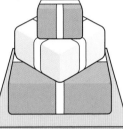

Calculate the surface area of the two larger cakes.

Exercise 26B

A **1** **A03** Triangle A is similar to triangle B.
The area of triangle A is 80 cm².
The area of triangle B is 500 cm².

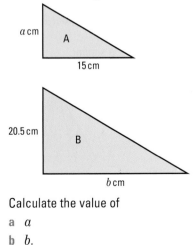

a cm   A
15 cm

20.5 cm   B
b cm

Calculate the value of
a a
b b.

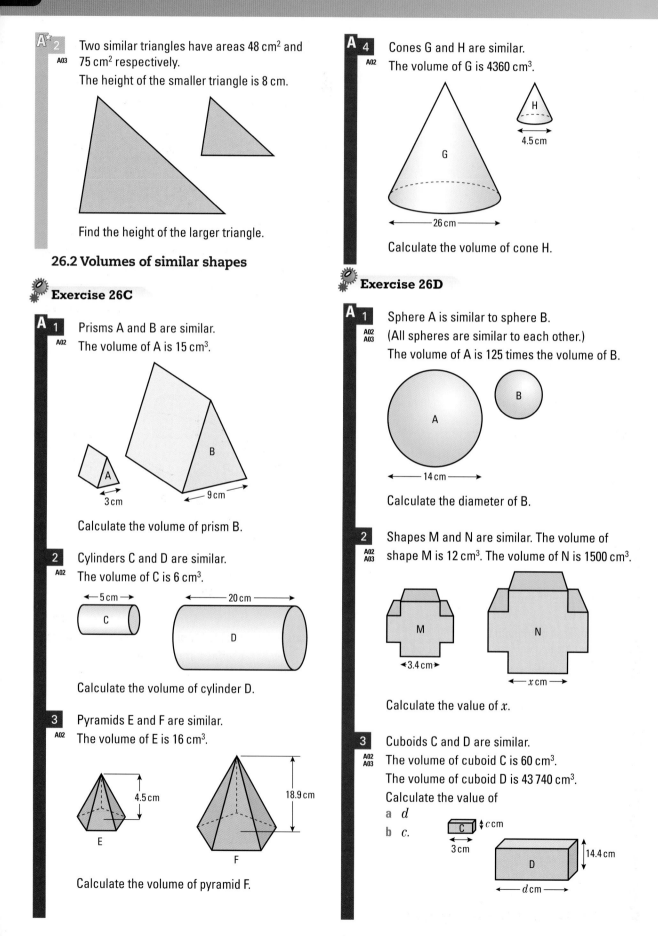

A⋆ **2**
A03

Two similar triangles have areas 48 cm² and 75 cm² respectively.

The height of the smaller triangle is 8 cm.

Find the height of the larger triangle.

26.2 Volumes of similar shapes

Exercise 26C

A 1
A02

Prisms A and B are similar.

The volume of A is 15 cm³.

Calculate the volume of prism B.

2
A02

Cylinders C and D are similar.

The volume of C is 6 cm³.

Calculate the volume of cylinder D.

3
A02

Pyramids E and F are similar.

The volume of E is 16 cm³.

Calculate the volume of pyramid F.

A 4
A02

Cones G and H are similar.

The volume of G is 4360 cm³.

Calculate the volume of cone H.

Exercise 26D

A 1
A02
A03

Sphere A is similar to sphere B.

(All spheres are similar to each other.)

The volume of A is 125 times the volume of B.

Calculate the diameter of B.

2
A02
A03

Shapes M and N are similar. The volume of shape M is 12 cm³. The volume of N is 1500 cm³.

Calculate the value of x.

3
A02
A03

Cuboids C and D are similar.

The volume of cuboid C is 60 cm³.

The volume of cuboid D is 43 740 cm³.

Calculate the value of

a d

b c.

A 4
A02 A03

Cylinders X and Y are similar.
The volume of X is $24\pi\,\text{cm}^3$.
The volume of Y is $375\pi\,\text{cm}^3$.

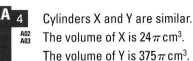

4.5 cm

Calculate the length of the radius of cylinder X.

A* 5
A02 A03

A deli sells many sizes of bagels. A large bagel has a diameter of 10 cm and a mass of 93.7 g. A mini bagel has a mass of 48 g.

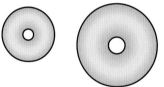

What is the diameter of the mini bagel?

26.3 Lengths, areas and volumes of similar shapes

 Exercise 26E

Give your answers to the following questions correct to 3 significant figures.

A 1
A02 A03

Cuboids S and T are similar. The surface area of cuboid S is 72 cm². The surface area of cuboid T is 1200 cm². The volume of cuboid S is 40 cm³.

Find the volume of cuboid T.

2
A02 A03

Cylinders V and W are similar. The volume of cylinder V is 1000 cm³. The volume of cylinder W is 2500 cm³. The area of the circular base of cylinder V is 30 cm².

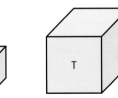

Find the area of the base of cylinder W.

A 3
A02 A03

A container has a surface area of 6000 cm² and a capacity of 12.4 litres.
Find the surface area of a similar container which has a capacity of 5.3 litres.

4
A02 A03

A water pistol can hold 800 cm³ of water. A similar larger water pistol can hold 12 500 cm³. The surface area of the water barrel on the larger pistol is 2400 cm².
Find the surface area of the water barrel on the smaller pistol.

A* 5
A02

A supermarket sells bottles of kitchen cleaner in three sizes of mathematically similar bottles: large (1 *l*), medium (500 m*l*), small (375 m*l*). The height of the large bottle is 20 cm and it has a label of area 64 cm².

a Calculate the height of:
  i the medium bottle
  ii the small bottle.
b Calculate the area of the label on:
  i the medium bottle
  ii the small bottle.

6
A02

A kitchen waste bin holds 40 *l* of refuse. A similar pedal bin holds 10 *l* of refuse. The kitchen waste bin has a surface area of 1.66 m².
What is the surface area of the pedal bin?

27 Proportion 2

Key Points

- **direct proportion:** occurs when a graph of two quantities is a straight line through the origin. This shows that the two quantities increase and decrease in the same ratio.
- **symbol ∝:** means 'is proportional to'.
- $y = kx$: the formula for direct proportion, where k is the constant of proportionality.

- **writing statements of proportionality:** use a constant of proportionality k.
  - 'y is proportional to the square of x' or $y \propto x^2$ means $y = kx^2$
  - 'y is proportional to the cube of x' or $y \propto x^3$ means $y = kx^3$
  - 'y is proportional to the square root of x' or $y \propto \sqrt{x}$ means $y = k\sqrt{x}$
  - 'y is inversely proportional to x' or $y \propto \dfrac{1}{x}$ means $y = \dfrac{k}{x}$

27.1 Direct proportion

Exercise 27A

Questions in this chapter are targeted at the grades indicated.

C 1 The table gives information about the variables x and H.

x	2	4	6	8	10
H	8	16	24	32	40

  a Plot the graph of H against x.
  b Is H directly proportional to x? Give a reason for your answer.

B 2 Here is a graph of $Y = kx$. Use the information in the graph to work out the value of k.

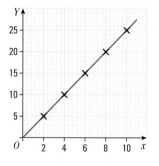

3 W is directly proportional to b.

  a Copy and complete this table of values for W and b.

b	5	10	15	25
W	12	24		

  b Given that $W = kb$, work out the value of k.

A 4 Y is directly proportional to q. $Y = 10$ when $q = 50$.

  a Sketch a graph of Y against q.
  b Work out a formula for Y in terms of q.
  c Use your formula to work out the value of Y when $q = 450$.

5 The cost (£C) of a tin of enamel paint is directly proportional to the volume (v ml) of paint in the tin. A tin containing 10 ml of paint costs £4.80.

  a Work out the cost of
     i 5 ml, **ii** 15 ml, **iii** 20 ml of the paint.
  b Work out a formula for C in terms of v.
  c The cost of a tin of enamel paint is £6.72. Work out the volume of paint in the tin.

27.2 Further direct proportion

Exercise 27B

B 1 g is directly proportional to x so that $g = kx$. $g = 5$ when $x = 37.5$. Work out the value of k.

2 Z is directly proportional to d so that $Z = kd$. $Z = 16.8$ when $d = 7$. Work out the value of k.

A 3 T is directly proportional to b. $T = 5.25$ when $b = 1.5$.

  a Show that $T = 3.5b$.
  b Work out the value of T when $b = 2.8$.
  c Work out the value of b when $T = 18.9$.

A 4
A02

The voltage V across a resistor (in volts) is directly proportional to the current c flowing through it (in amps).

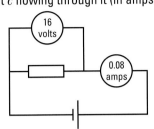

a Show that $V = 200c$.

b Work out the value of c when $V = 11$ volts.

5
A02

The amount of stretch S of a spring (in mm) is directly proportional to the mass m on the spring (in grams).

a Find a formula for S in terms of m.

b Find the amount of stretch of the spring when the mass is 650 g.

c Find the mass that will stretch the spring by 74.5 mm.

45 mm

360 g

6
A02

A manufacturer produces rectangular rubber mats in a variety of sizes. The length (L) of each mat is directly proportional to its width (w).
A mat with a width of 24 cm has a length of 64 cm.
Work out the length of a mat with width 33 cm.

27.3 Writing statements of proportionality and formulae

Exercise 27C

A 1

Write down i the statement of proportionality, ii the formula, for each of the following.
Use the symbol $\propto$.

a A is directly proportional to b.

b L is proportional to the square root of m.

c S is proportional to the square of t.

d R is proportional to the cube of p.

e Y is proportional to the cube root of z.

f P is inversely proportional to q.

g G is inversely proportional to the square of h.

h V is inversely proportional to the square root of m.

i F is inversely proportional to the cube of d.

j W is inversely proportional to the cube root of t.

A 2

The number of birds' eggs in an incubator, E, is proportional to the square root of the length, l cm, of the incubator.

a Write down the statement of proportionality.

b Write down a formula for E in terms of l and k (the constant of proportionality).

3
A02

The length, L mm, of a snake is proportional to the square of the circumference, c mm, of the snake. Write down a formula for L in terms of c.

4

The amount of honey, H ml, produced by a hive of bees is proportional to the cube root of the number of bees, b, in the hive.
Write down a formula for H in terms of b.

5

A formula for the temperature (T°C) in a furnace is given by $T = kf^3$, where f is the amount of fuel burned (in kilograms) and k is a constant.
What is the relationship between T and f?

6

A formula for the number of revolutions (R) of a wheel is given by $R = \dfrac{k}{\sqrt{c}}$, where c is the circumference of the wheel and k is a constant.
Alexandra says that R is proportional to the square root of c.
Is Alexandra correct?
Give a reason for your answer.

27.4 Problems involving square and cubic proportionality

Exercise 27D

A 1

Y is proportional to the square of a so that $Y = ka^2$.
$Y = 24$ when $a = 6$. Work out the value of k.

2

H is proportional to the cube of j so that $H = kj^3$.
$H = 90$ when $j = 2.5$. Work out the value of k.

3

$X \propto c^2$. $X = 54$ when $c = 4$.
Find a formula for X in terms of c.

4

$T \propto m^3$. $T = 150$ when $m = 5$.
Find a formula for T in terms of m.

5

G is proportional to the cube of y. $G = 10$ when $y = 2$. Work out the value of G when $y = 4$.

A **6** A tennis ball is thrown vertically upwards with a speed of v m/s. The height, H metres, reached by the ball is proportional to the square of v. When $v = 30$ m/s, $H = 30$ m. Work out the value of:

  a H when $v = 20$ m/s
  b v when $H = 15$ m.

7 The surface area, A mm², of a petal is proportional to the square of its length, l mm.
A02 When the length of a petal is 12 mm its surface area is 44 mm². Work out the length of a petal that has a surface area of 85 mm².
Give your answer to 2 decimal places.

8 This table gives the mass and length of some toy snakes. The mass (g) of each snake is
A02 proportional to the square of its length (cm). Copy and complete this table.

Mass (grams)	Length (cm)
25	5
38.44	
54	
	8.8

27.5 Problems involving inverse proportion

Exercise 27E

A **1** P is inversely proportional to x so that $P = \dfrac{k}{x}$.
$P = 2.75$ when $x = 25$. Work out the value of k.

2 $F \propto \dfrac{1}{g}$
$F = 4.25$ when $g = 16$.

  a Work out the value of F when $g = 18$.
  b Work out the value of g when $F = 17$.

A **3** $N \propto \dfrac{1}{t}$
$N = 3.8$ when $t = 5$.

  a Work out the value of N when $t = 9$.
  b Work out the value of t when $N = 9.8$.

4 The volume V (m³) of a gas is inversely proportional to the pressure P (N/m²).
$V = 6$ m³ when $P = 750$ N/m².
Work out the volume of the gas when the pressure is 1000 N/m².

5 H is inversely proportional to the square of v so that $H = \dfrac{k}{v^2}$. $H = 2.75$ when $v = 4$.

  a Work out the value of k.
  b Work out the value of H when $v = 5$.
  c Work out the value of v when $H = 9$.

6 $Q \propto \dfrac{1}{s^2}$
$Q = 0.725$ when $s = 6$.

  a Work out the value of Q when $s = 4.5$.
  b Work out the value of s when $Q = 2$.

7 The speed (S), in revolutions per hour, at
A02 which each cog in a machine turns is inversely proportional to the square of the radius (r), in cm, of the cog. When $r = 6$ cm, $S = 95$ revolutions per hour.

  a Find a formula for S in terms of r.
  b Work out the value of S when $r = 5$.

28 Probability

🔍 Key Points

- **probability (P):** how likely it is an event will occur.
  - $0 \leqslant P \leqslant 1$
  - for an event which is certain P = 1
  - for an event which is impossible P = 0
  - can be written as a fraction, decimal or percentage
- **sample space:** all the possible outcomes of one or more events. Can be presented in a sample space diagram.
- **mutually exclusive outcomes:** events that cannot happen at the same time.
  - for mutually exclusive events A and B, P(A or B) = P(A) + P(B)
  - P(A or B or C or ...) = P(A) + P(B) + P(C) + ...
  - for mutually exclusive events A and not A, P(not A) = 1 − P(A)
  - If a set of mutually exclusive events contains all possible outcomes, then the sum of their probabilities must come to 1.
- **relative frequency:** the frequency of an observed value divided by the total number of observations.

- **independent events:** two events whereby one event does not affect the other event.
  - for two independent events A and B, P(A and B) = P(A) × P(B)
  - P(A and B and C and ...) = P(A) × P(B) × P(C) × ...
- **probability tree diagram:** a diagram used to solve problems involving a series of successive events. All of the possible outcomes can be seen by following the paths along branches of the tree.
- **conditional probability:** when one outcome affects another outcome, so that the probability of the second outcome depends on what has already happened in the first outcome.
- **finding the probability of equally likely outcomes:** use the formula
  $$\text{probability} = \frac{\text{number of successful outcomes}}{\text{total number of possible outcomes}}$$
- **calculating the estimated probability:** use the formula
  $$\text{estimated probability} = \frac{\text{number of successful trials}}{\text{total number of trials}}$$

28.1 Writing probabilities as numbers

⚙ Exercise 28A

Questions in this chapter are targeted at the grades indicated.

1 Rio spins this 5-sided spinner. The spinner is fair. Work out the probability that the spinner will land on

  a black   b white   c grey.

2 Work out the probability of each of the following.
  a rolling the number 5 with an ordinary dice
  b rolling a number less than 4 with an ordinary dice
  c taking a red queen from an ordinary pack of cards
  d taking a black card from an ordinary pack of cards
  e taking a picture card (jack, queen or king) from an ordinary pack of cards.

3 A bag contains 4 spotted discs and 7 striped discs. A disc is taken at random from the bag. Work out the probability that the disc will be:
  a spotted
  b striped
  c checked.

4 The faces of a 10-sided dice are numbered from 1 to 10.
Work out the probability of rolling each of the following.
  a an even number   b a 5 or a 6
  c a multiple of 3   d a prime number
  e a factor of 12.

5 A local hospice runs a raffle. It sells 3200 tickets in total. Harris buys 4 tickets in the raffle. Work out the probability that he will win the raffle.

D **6** A packet of fruit teas contains 4 blackcurrant teabags, 5 lemon teabags and 3 elderflower teabags. One of these teabags is taken from the packet at random.
Work out the probability that the teabag will be:
a lemon
b blackcurrant
c elderflower
d lemon or blackcurrant.

7 A letter is chosen at random from the word MATHEMATICS.
Write down the probability that it will be:
a T   b E   c A or C
d a consonant   e L.

8 The table gives the numbers of boys and the numbers of girls in a nursery school and whether they can write their name.

	Can write name	Cannot write name	Total
Boys	10	16	
Girls	13	17	
Total			56

a Copy and complete the table.
b One of these children is chosen at random.
Use the information in your table to work out the probability that the child will be:
i able to write his or her name
ii a girl
iii a boy who cannot write his name.

9
A03 The pie chart gives information about how some people travelled from Oxford to Leicester one day. One person is picked at random.

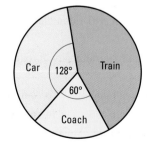

Use the information in the pie chart to work out the probability that the person:
a travelled by car
b travelled by train.

A **10**
A03 A group of friends go to a restaurant. 60% are women. 45% prefer to eat from the buffet. 20% are men who prefer to have waiter service. One of these friends is chosen at random.
Work out the probability that the friend is a woman who prefers to eat from the buffet.

Exercise 28B

C **1** Two ordinary dice are rolled.
Use a sample space diagram to work out the probability of getting each of the following outcomes.
a a total score of
i 4   ii 7   iii 6 or less
b an odd number on each dice
c a difference of 3 between the numbers on the two dice.

2 A fair 5-sided spinner and a fair coin are spun.
a Copy and complete the following sample space diagram to show all the possible outcomes.

Coin
$$\begin{array}{c|ccccc} \text{H} & (1,\text{H}) & & & & \\ \text{T} & (1,\text{T}) & (2,\text{T}) & & & \\ \hline & 1 & 2 & 3 & 4 & 5 \\ & & & \text{Spinner} & & \end{array}$$

b Work out the probability of getting:
i a 2 on the spinner and a tail on the coin
ii a number less than 4 on the spinner
iii a number greater than 2 on the spinner and a head.

3 Two fair 3-sided spinners are spun and the product of the numbers is calculated.
a Copy and complete this sample space diagram to show all the possible outcomes.

		Spinner A	
	1	**2**	**3**
1	1	2	
2		4	
3			

Spinner B

b Work out the probability of getting a product:
i greater than 7   ii of 4   iii of 6.

B **4** A fair 5-sided spinner and a fair 3-sided spinner are spun.

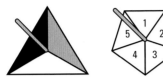

a Draw a sample space diagram to show all the possible outcomes.

b Use your sample space diagram to work out the probability of the spinners landing on:

  i grey and an even number

  ii white and an odd number

  iii black and any number.

5 Naomi has two sections in her cutlery drawer in which she puts coloured spoons and forks for feeding her baby. In section S, she has a green spoon, a blue spoon, a yellow spoon and an orange spoon. In section F, she has matching coloured forks to the spoons in section S. Naomi takes a spoon and a fork at random from the drawer.

a Draw a sample space diagram to show all the possible outcomes.

b Work out the probability that the spoon and fork will be:

  i both orange

  ii the same colour

  iii different colours.

A **6** A golf club has an annual competition in which both men and women can compete. Each player in the competition plays every other player. Four men and three women sign up to take part. Two players are picked at random to play first. Work out the probability that the first game will be played by a man and a woman.

A☆ **7** A fair 8-sided spinner is spun and an ordinary 6-sided dice is rolled. The numbers the spinner lands on are used for x-coordinates, and the numbers the dice lands on are used for y-coordinates.

A03

Find the probability that the point generated by the numbers the spinner and dice land on lies on each of the following lines.

  a $y = 2$ b $x + y = 6$ c $y = x + 3$

  d $y = 3x - 2$ e $y = \frac{1}{3}x + 2$

28.2 Mutually exclusive outcomes

Exercise 28C

1 Here is a number of shapes.

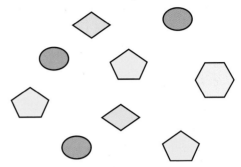

One of these shapes is chosen at random. Work out the probability that the shape will be:

a a pentagon

b a rhombus

c a hexagon or an oval.

D **2** A bag contains some marbles. The colour of each marble is either orange, green or yellow. A marble is taken at random from the bag. The probability that the marble will be orange is 0.4. The probability that the marble will be green is 0.3.
Work out the probability that the marble will be orange or green.

3 The table gives the probability of getting each of 0, 1, 2, 3 and 4 on a biased 5-sided spinner.

Number	0	1	2	3	4
Probability	0.1	0.15	0.25	0.3	0.2

Work out the probability of getting:

a 0 or 2   b 1 or 3

c 1 or 2   d 2 or 3 or 4

4 Below is a number of lettered cards.

| Q | U | A | D | R | I | L | A | T | E | R | A | L |

One of these cards is selected at random. Work out the probability of getting:

a an A   b an L

c an I   d an A or an R

e an A or an I   f an L or a D.

D 5   Below are some cards with shapes drawn on them.

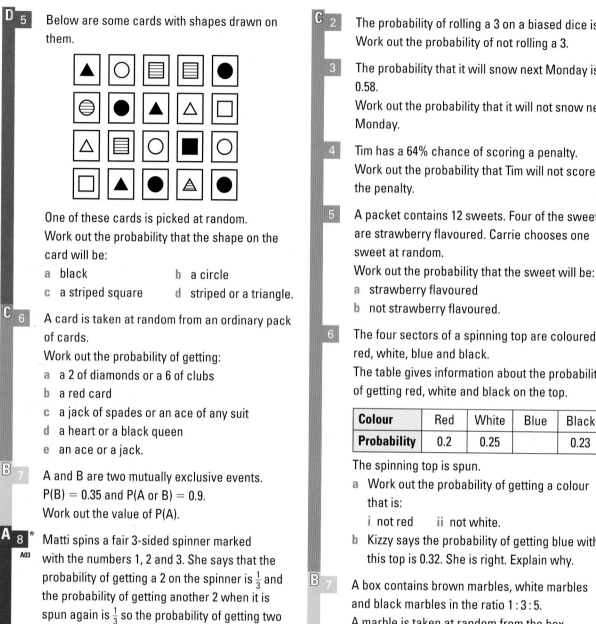

One of these cards is picked at random.
Work out the probability that the shape on the card will be:

a   black   b   a circle

c   a striped square   d   striped or a triangle.

C 6   A card is taken at random from an ordinary pack of cards.
Work out the probability of getting:

a   a 2 of diamonds or a 6 of clubs

b   a red card

c   a jack of spades or an ace of any suit

d   a heart or a black queen

e   an ace or a jack.

B 7   A and B are two mutually exclusive events.
P(B) = 0.35 and P(A or B) = 0.9.
Work out the value of P(A).

A 8*   Matti spins a fair 3-sided spinner marked
A03   with the numbers 1, 2 and 3. She says that the probability of getting a 2 on the spinner is $\frac{1}{3}$ and the probability of getting another 2 when it is spun again is $\frac{1}{3}$ so the probability of getting two 2s is $\frac{1}{3} + \frac{1}{3} = \frac{2}{3}$. Is Matti correct? Explain why.

Exercise 28D

D 1   The pie chart shows the proportion of people choosing a puppy from three different breeds of dog at a kennels: labrador, spaniel, terrier.

One customer is asked at random which breed of puppy they chose.
Work out the probability
that they chose a terrier.

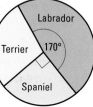

C 2   The probability of rolling a 3 on a biased dice is $\frac{1}{4}$.
Work out the probability of not rolling a 3.

3   The probability that it will snow next Monday is 0.58.
Work out the probability that it will not snow next Monday.

4   Tim has a 64% chance of scoring a penalty.
Work out the probability that Tim will not score the penalty.

5   A packet contains 12 sweets. Four of the sweets are strawberry flavoured. Carrie chooses one sweet at random.
Work out the probability that the sweet will be:

a   strawberry flavoured

b   not strawberry flavoured.

6   The four sectors of a spinning top are coloured red, white, blue and black.
The table gives information about the probability of getting red, white and black on the top.

Colour	Red	White	Blue	Black
Probability	0.2	0.25		0.23

The spinning top is spun.

a   Work out the probability of getting a colour that is:

    i   not red   ii   not white.

b   Kizzy says the probability of getting blue with this top is 0.32. She is right. Explain why.

B 7   A box contains brown marbles, white marbles and black marbles in the ratio 1 : 3 : 5.
A marble is taken at random from the box.
Work out the probability that the marble will be:

a   brown   b   not brown   c   not black.

A 8   For two mutually exclusive events X and Y,
A03   P(X) = 0.5 and P(X or Y) = 0.9.
Work out P(not Y).

28.3 Estimating probability from relative frequency

Exercise 28E

1   Miriam rolls a dice 200 times and gets 30 sixes.
Work out an estimate for the probability of getting a six on Miriam's dice.

2 Out of 40 tomato plants in a greenhouse, 34 produce more than 10 tomatoes each.
Work out an estimate for the probability that a tomato plant in the greenhouse will produce more than 10 tomatoes.

D 3 The sectors of a 5-sided spinner are each coloured red, white, black, blue or orange.
The table gives the results when the spinner is spun 360 times.

Colour	Red	White	Black	Blue	Orange
Frequency	48	60	132	84	36

a Use the information in the table to find an estimate for the probability of getting black.

b Is this a fair spinner?
Give a reason for your answer.

4 Shuffle a pack of cards and choose a card at random. Record the card and then replace it in the pack. Repeat this 100 times.

a Use your results to find an estimate for the probability of picking a king from the pack.

b How could you improve on your answer to part a?

5 Jake records the number of 3s he gets when he spins a 4-sided spinner 20, 200 and 2000 times.
The table below shows his results.

Number of spins	20	200	2000
Number of 3s	4	44	490

Use this information to work out the best estimate for getting a 3 on Jake's spinner.
Give a reason for your answer.

A 6 * Mia says that when she drops a drawing pin it lands on its head more times than its tail.

Head   Tail

Carry out an experiment to find an estimate for the probability that a drawing pin will land on its head.
Explain all stages of your work.

28.4 Finding the expected number of outcomes

Exercise 28F

1 Lily rolls a fair dice 120 times.
How many 5s can she expect to get?

2 Mike spins an ordinary coin 80 times.
How many tails can he expect to get?

D 3 The table gives the probability of spinning the numbers 1, 2, 3, 4 and 5 on a 5-sided spinner.

Number	1	2	3	4	5
Probability	0.15	0.2	0.3	0.1	0.25

Saleem spins the spinner 250 times.
Work out an estimate for the number of 2s that Saleem will get.

B 4 A bag contains lettered tiles. There are 2 Ts, 5 Ms and 3 Ls.
A tile is taken at random from the bag and then replaced. This is done 200 times.
Copy and complete the table to show the expected numbers of Ts, Ms and Ls that will be taken from the bag.

Letter	T	M	L
Expected number			

5
A02 Farah rolls two ordinary dice and records the result. She does this 180 times. One possible outcome is two 6s. Find an estimate for the number of times Farah will get two 6s.

6 * The probability of picking out a premium prize in a lucky dip is $\frac{1}{40}$. Darcy says that if she has 80 dips, she should win two prizes. Is she correct?
Give a reason for your answer.

A 7 A card is taken from an ordinary pack of cards. It is then replaced. If this is done 520 times, work out the expected number of the following that will be taken from the pack.

a red kings

b picture cards (jacks, queens or kings)

c queens or even numbers.

A 8 * A dentist estimates that the probability a patient will come to her surgery about a lost filling is 0.375.

Of the next 160 patients who come to the surgery, 50 have a lost filling.

How accurate is the dentist's estimate of this probability? Explain your answer.

28.5 Independent events

Exercise 28G

C 1 Mei Yih and Dave play two different card games.
The probability that Mei Yih will win the first game is 0.6.
The probability that she will win the second game is 0.3.
Work out the probability that Mei Yih will win both games.

2 The probability that Adam will be picked for the rugby team is $\frac{3}{5}$.
The probability that he will remember his sports kit is $\frac{5}{6}$.
Work out the probability that Adam will be picked for the team and remember his kit.

3 The probability that a newspaper boy will deliver a newspaper to Stuart's house today is 0.9.
The probability that Stuart has toast for breakfast is 0.7.
Work out the probability that the newspaper boy will deliver a newspaper to Stuart's house today and he has toast for breakfast.

4 The probability that Sam will remember to take his glasses to school is 0.75.
The probability that he will remember to take his lunchbox is 0.64.
Work out the probability that Sam will:
a forget his glasses
b forget his lunchbox
c forget both.

B 5 A card is taken at random from each of two ordinary packs of cards, pack X and pack Y.
Work out the probability of getting:
a a black card from pack X and a red card from pack Y
b a heart from pack X and a red card from pack Y
c a queen from pack X and an 8 or a 9 from pack Y
d a jack from pack X and a red jack from pack Y
e a king of spades or an ace of spades from each pack.

A 6 Simon and Susannah are sorting coins for a parking meter. The table below gives information about these coins.

	Number of coins			
	10p	**20p**	**50p**	**£1**
Simon	3	2	4	2
Susannah	2	3	1	1

Simon and Susannah each pick one of their own coins at random.
Work out the probability that:
a Simon picks a 20p coin and Susannah picks a £1 coin
b neither of them picks a silver coin
c neither of them picks a 50p coin.

28.6 Probability tree diagrams

Exercise 28H

A 1 Bag P contains 3 yellow balls and 5 green balls.
Bag Q contains 4 yellow balls and 1 green ball.
A ball is taken at random from each bag.
a Copy and complete the tree diagram to show all the possible outcomes.

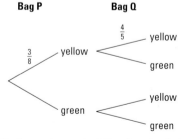

b Work out the probability that the balls will both be:
i green
ii yellow
iii different colours.

A **2** A box of chalk contains 4 white chalks and 8 pink chalks. A piece of chalk is taken at random from the box and then returned.
A second piece of chalk is now taken at random from the box and then returned.

a Copy and complete the tree diagram to show all the possible outcomes.

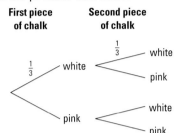

b Work out the probability that only one piece of chalk picked will be white.

3 Kafi and Wayne each have a bag of jelly babies. In Kafi's bag there are 4 yellow sweets and 6 green sweets.
Wayne's bag contains 3 yellow sweets and 4 green sweets.
Kafi and Wayne each take a sweet at random from their own bag.

a Draw a tree diagram to show all the possible outcomes.

b Use your tree diagram to work out the probability that the sweets will:
 i both be yellow
 ii each have the same colour.

4 Paul has to catch two trains to get to work. The probability that the first train will be late is 0.3.
The probability that the second train will be late is 0.2.

a Work out the probability that:
 i the first train will not be late
 ii the second train will not be late.

b Draw a tree diagram to show all the possible outcomes.

c Work out the probability that:
 i both trains will be late
 ii only one train will be late
 iii neither train will be late.

A **5** Charlie rolls a red dice and a blue dice. Both dice are biased. The probability of getting a 6 on the red dice is 0.2. The probability of getting a 6 on the blue dice is 0.35.

a Draw a tree diagram to show all the possible outcomes.

b Work out the probability of getting a 6 on:
 i both dice
 ii only one dice
 iii the red dice only.

6
A02
A03
The probability that Barry will get to work early on Thursday is $\frac{4}{5}$. The probability that Barry will get to work early on Friday is $\frac{5}{8}$.
Work out the probability that he will be early on at least one of these days

7 A card is taken at random from an ordinary pack of cards. It is then replaced.
Another card is now taken at random from the pack of cards.
Work out the probability of the following.

a Both cards are picture cards (jacks, queens or kings).

b Neither of the cards is a diamond.

c Only one of the cards is an even number.

d Only one of the cards is a red jack.

e At least one of the cards is a queen.

8
A02
Three drawing pins are dropped.
They can either land pin up or pin down, with equal probability.

a Show all possible outcomes.

b Work out the probability of the drawing pins landing with:
 i 3 pins up
 ii 2 pins up and 1 pin down (in any order).

9
A02
Alex's toaster is not working very well.
The probability that his toast will burn is 0.2.
He makes three pieces of toast.
What is the probability that one of them will burn?

A* **10**
A03
There is an 88% chance that a Dumbo battery is not faulty. Dean orders 300 triple packs of Dumbo batteries for his shop.
Find an estimate for the number of packs that contain exactly 2 faulty batteries.

28.7 Conditional probability

Exercise 28I

1 A box of chocolates contains 8 plain chocolates and 3 milk chocolates.
A chocolate is taken at random from the box and eaten. A second chocolate is now taken at random from the box.

a Copy and complete the tree diagram to show all the possible outcomes.

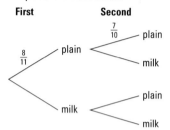

First   **Second**

b Work out the probability that the chocolates will:

  i both be plain

  ii be the same type of chocolate

  iii be one of each type of chocolate.

2 6 boys and 4 girls want to represent the school at a local council function.
Two of these students are picked at random.

a Copy and complete the tree diagram to show all the possible outcomes.

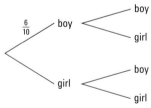

First   **Second**

b Work out the probability of getting:

  i 2 girls

  ii 1 girl

  iii at least one boy.

3  Nine letter cards marked A–I are shuffled thoroughly. The top two cards are turned face up on a table. Draw a probability tree diagram and use it to work out the probability that:

a the letters will both be vowels

b at least one letter will be a vowel.

4 The Rich family are going on holiday.
The probability that their taxi will arrive to pick them up on time is $\frac{7}{8}$. If the taxi arrives on time, the probability that they will miss their flight is $\frac{1}{10}$. If their taxi is late, the probability that they will miss their flight is $\frac{3}{5}$. Use a tree diagram to work out the probability that:

a the taxi is late, but the Rich's still catch their flight

b the Rich's miss their flight.

5 Anya travels to work by car or by bicycle.
The probability that she travels by car is 0.35.
If she travels to work by car, the probability that she will be late is 0.12. If she travels to work by bicycle, the probability that she will be late is 0.25.
Work out the probability that she will not be late.

6 A store imports two types of digital photo frames, P and Q, in equal numbers. The probability that type P will be faulty is 0.18. The probability that type Q will be faulty is 0.24. If type P is faulty, the probability that it can be repaired is 0.36. If type Q is faulty, the probability that it can be repaired is 0.45. A frame is picked at random.
Work out the probability that the frame is faulty and can be repaired.
Give your answer in standard form.

7 Lee has 12 coins in his pocket. They are all €2 coins, but 5 are from France and 3 are from Spain. He picks two coins from his pocket.
Work out the probability that the coins will be:

a both from France

b not from Spain.

8 Two cards are taken at random from an ordinary pack of cards.
Work out the probability that the cards will be:

a both picture cards (kings, queens or jacks)

b both red

c from same suit.

9 The probability that it will snow today is $\frac{2}{5}$.
If it does not snow today, the probability that it will snow tomorrow is $\frac{3}{5}$.
Work out the probability that it will snow today or tomorrow.

A* 10

A02
A03

Gemma is taking a practical biology exam.
She will be shown three samples A, B and C, but
does not know which she will have.
The probability that she will get A, or B, or C is
0.35, or 0.42, or 0.23 respectively. The probability
that Gemma will not pass the exam with each
sample is 0.24, or 0.36, or 0.15 respectively.
Work out the probability that Gemma will pass
the exam.

29 Pythagoras' Theorem and trigonometry 2

Key Points

⊛ **graph of $y = \sin\theta°$:**

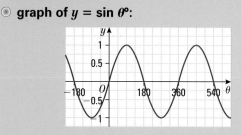

- ⊛ cuts the θ-axis at …, $-180, 0, 180, 360, 540, \ldots$
- ⊛ has a period of $360°$
- ⊛ maximum value of 1 at $\theta° = \ldots, 90, 450, \ldots$
- ⊛ minimum value of -1 at $\theta° = \ldots, -90, 270, \ldots$

⊛ **graph of $y = \cos\theta°$:**

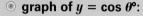

- ⊛ cuts the θ-axis at …, $-90, 90, 270, 450, \ldots$
- ⊛ has a period of $360°$
- ⊛ maximum value of 1 at $\theta° = \ldots, 0, 360, \ldots$
- ⊛ minimum value of -1 at $\theta° = \ldots, -180, 180, \ldots$

⊛ **the sine rule:** a rule that can be used with any triangle when the problem involves 2 sides and 2 angles.

- ⊛ $\dfrac{a}{\sin A} = \dfrac{b}{\sin B} = \dfrac{c}{\sin C}$
- ⊛ $\dfrac{\sin A}{a} = \dfrac{\sin B}{b} = \dfrac{\sin C}{c}$

⊛ **the cosine rule:** a rule that can be used with any triangle when the problem involves 3 sides and 1 angle.

⊛ $a^2 = b^2 + c^2 - 2bc \cos A$
$b^2 = a^2 + c^2 - 2ac \cos B$
$c^2 = a^2 + b^2 - 2ab \cos C$

⊛ $\cos A = \dfrac{b^2 + c^2 - a^2}{2bc}$

$\cos B = \dfrac{a^2 + c^2 - b^2}{2ac}$

$\cos C = \dfrac{a^2 + b^2 - c^2}{2ab}$

⊛ **finding the length of the longest diagonal of a cuboid with dimensions a, b, c:** use the formula $d = \sqrt{a^2 + b^2 + c^2}$.

⊛ **using quadrants to determine if $\sin\theta$ and $\cos\theta$ is positive or negative:** use the following diagram where the angle θ is the angle a line moving anticlockwise from the positive x-axis makes with the x-axis.

$\sin+$ $\cos-$	$\sin+$ $\cos+$
2nd	**1st**
3rd	**4th**
$\sin-$ $\cos-$	$\sin-$ $\cos+$

⊛ **finding the area of a triangle using sine:** use the formula:

area of ABC $= \frac{1}{2}ab \sin C = \frac{1}{2}bc \sin A = \frac{1}{2}ac \sin B$

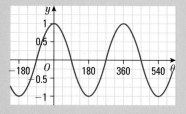

29.1 Pythagoras' Theorem and trigonometry in three dimensions

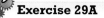

Exercise 29A

Questions in this chapter are targeted at the grades indicated.

Where necessary give lengths correct to 3 significant figures and angles correct to one decimal place.

A **1** ABCDEF is a triangular prism. In triangle ABC, angle CAB $= 90°$, AB $= 5$ cm and AC $= 12$ cm. In rectangle ABED, the length of BE $= 15$ cm.

  a Calculate the length of CB.

  b Calculate the length of
   i CE
   ii AF.

  c Calculate the size of
   i angle FED
   ii angle FAD.

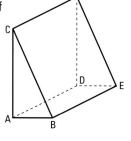

2 For the cuboid ABCDEFGH, show that
BH² = AB² + AD² + DH².

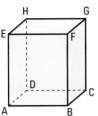

3 Dave wants to put the remains of a packet of breadsticks in a tin. The tin is in the shape of a cuboid which is 22 cm long, 9 cm wide and 8 cm deep. Will a breadstick, of length 25 cm, fit in the tin without breaking?

4 A decorator uses a stick of 35 cm length to stir his paint. When it is not in use, he rests it in an empty paint tin. The tin is a cylinder of radius 9 cm and height 20 cm. The stick rests so that one end is against the circumference of the base as shown.
How much of the stick is outside the tin?

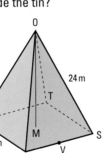

5 The diagram shows the top of a clock tower in the shape of a square-based pyramid. The sides of the square base are 16 m. The top vertex is above the centre of the square. The centre of the front of the tower is at V. The slant height of the pyramid is 24 m.
Calculate:

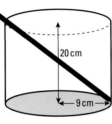

a i the distance from one corner to the opposite corner (e.g. SU)
   ii the distance from any corner to the centre of the base (M).
b the vertical height (OM)
c the angle between a slanted edge and the base (e.g. ∠OUS)
d the angle between two opposite slanted edges (e.g. ∠UOS)
e the distance from the top vertex (O) to the midpoint of the front side (V)
f the angle between a slanted edge and the side of the base (e.g. ∠OSV).

29.2 Angle between a line and a plane

Exercise 29B

Where necessary give lengths correct to 3 significant figures and angles correct to one decimal place.

1 The diagram shows a pyramid.
The base, ABCD, is a horizontal rectangle in which AB = 12 cm and AD = 9 cm. The vertex, O, is vertically above the centre of the base and OA = 27 cm.

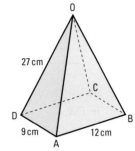

Calculate the size of the angle that OA makes with the horizontal plane.

2 ABCDEF is a triangular prism. In triangle ABC, angle CAB = 90°, AB = 8 cm and AC = 10 cm. In rectangle ABED, the length of BE = 5 cm. Calculate the size of the angle between:
a CB and ABED
b CD and ABED
c CE and ABED
d BC and ADFC.

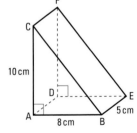

3 The diagram shows a square-based pyramid. The lengths of sides of the square base, ABCD, are 8 cm and the base is on a horizontal plane. The centre of the base is the point M and the vertex of the pyramid is O, so that OM is vertical. The point E is the midpoint of the side AB. OA = OB = OC = OD = 20 cm.

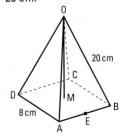

Calculate the size of the angle between OE and the base ABCD.

A* 4
A02 A03

WXYZ is a rectangular fruit garden and PY is a vertical pole. The gardener fixes protective netting, supported by strings, from the top of the pole (P) to the corners of the fruit garden.

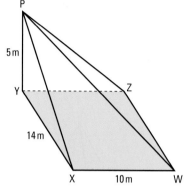

a Calculate the lengths of the strings:
   i PZ   ii PW   iii PX
b Calculate the angles made by the strings
   i PZ   ii PW   iii PX
   with the horizontal.

5
A02 A03

These are two different shapes in a bag of children's bricks. Shape 1 is a square-based pyramid with each side of the base 10 cm and each slanted edge 13 cm.

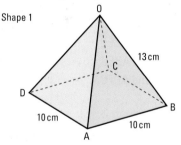

Shape 2 is a cuboid with each side of the base 10 cm and height 8 cm.

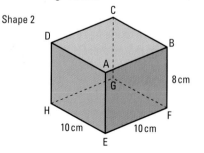

Jacob puts the pyramid exactly on top of the cuboid on a horizontal floor.
a How high is the top vertex (O) above the floor?
b What is the angle between OE and a diagonal of the base of the cuboid?

A* 6
A02 A03

This diagram shows a cube of cheese of edge 2 cm. A corner of the cube is cut off as shown with A, B, C the midpoints of the edges.

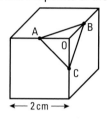

a In the pyramid OABC, find:
   i ∠AOB   ii ∠ACB.
b What is the volume of cheese remaining after the corner has been cut off?

29.3 Trigonometric ratios for any angle

Exercise 29C

Give angles correct to one decimal place.

A 1 For $-540 \leqslant \theta \leqslant 180$, sketch the graph of
   a $y = \sin \theta°$ b $y = \cos \theta°$.

A* 2 Find all values of θ in the interval -540 to 180 for which
   a $\sin \theta° = 0.6$ b $\cos \theta° = 0.2$.

3 a Show that one solution of the equation $4\sin \theta° = 1$ is 14.5, correct to 1 decimal place.
   b Hence solve the equation $4\sin \theta° = 1$ for values of θ in the interval 0 to 720.

4 a Show that one solution of the equation $5\cos \theta° = -3$ is 126.9 correct to 1 decimal place.
   b Hence find all values of θ in the interval -360 to 360 for which $5\cos \theta° = -3$.

29.4 Finding the area of a triangle using $\frac{1}{2}ab \sin C$

Exercise 29D

Give lengths and areas correct to three significant figures and angles correct to one decimal place.

A 1 Work out the area of each of these triangles.

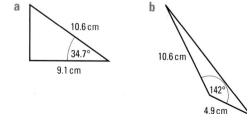

A

c

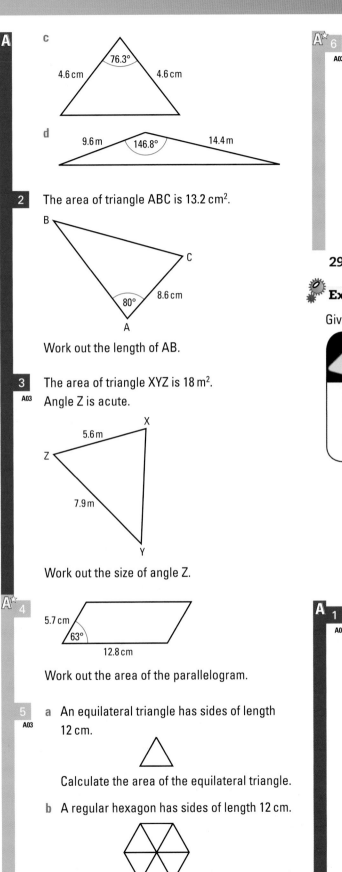

76.3°
4.6 cm   4.6 cm

d

9.6 m   146.8°   14.4 m

2 The area of triangle ABC is 13.2 cm².

B

C

8.6 cm
80°

A

Work out the length of AB.

3 The area of triangle XYZ is 18 m².
A03 Angle Z is acute.

X
5.6 m
Z

7.9 m

Y

Work out the size of angle Z.

A★ **4**

5.7 cm
63°
12.8 cm

Work out the area of the parallelogram.

5 **a** An equilateral triangle has sides of length
A03   12 cm.

Calculate the area of the equilateral triangle.

b A regular hexagon has sides of length 12 cm.

Calculate the area of the regular hexagon.

A★ **6** The diagram shows a sector, OXY, of a circle,
A03 centre O. The radius of the circle is 6.5 cm and
the size of angle XOY is 54°.

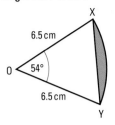

X
6.5 cm
O   54°
6.5 cm
Y

Work out the area of the segment of the circle
shown shaded in the diagram.

29.5 The sine rule

Exercise 29E

Give lengths correct to three significant figures.

Results**Plus**
Watch Out!

You can only use Pythagoras'
Theorem in a right-angled
triangle.

Results**Plus**
Exam Tip

Check that your answer is sensible:
the greater length is always
opposite the greater angle.

A **1** Find the lengths of the sides marked with letters
A02 in these triangles.

a

a   4.3 cm
29°   54°

b

62°

b

72°
7.2 cm

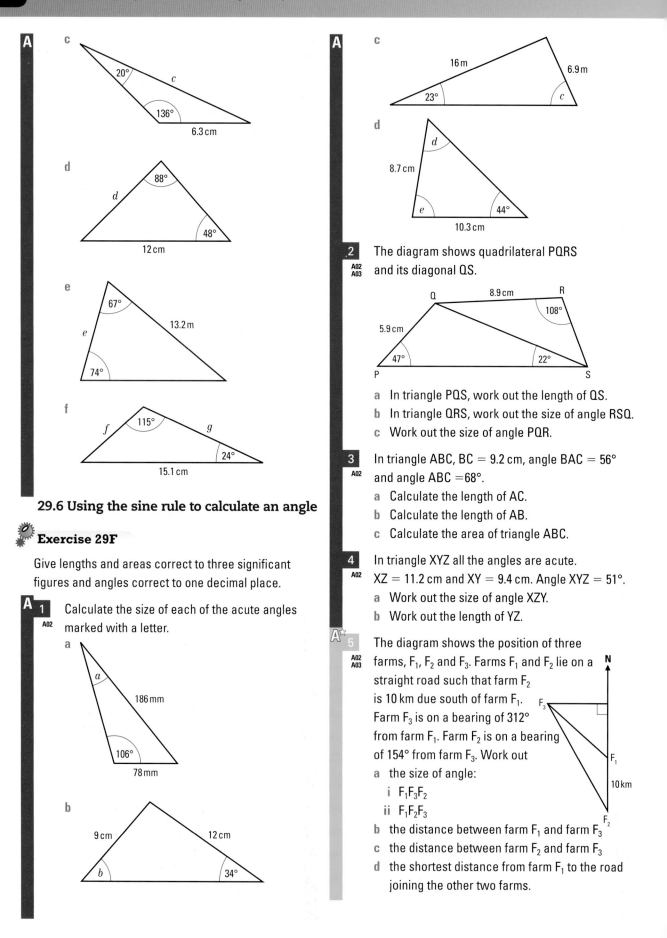

A

c

20°

c

136°

6.3 cm

d

88°

d

12 cm

48°

e

67°

13.2 m

e

74°

f

115°

f

g

24°

15.1 cm

29.6 Using the sine rule to calculate an angle

Exercise 29F

Give lengths and areas correct to three significant figures and angles correct to one decimal place.

A 1
A02

Calculate the size of each of the acute angles marked with a letter.

a

a

186 mm

106°

78 mm

b

9 cm

12 cm

b

34°

A

c

16 m

6.9 m

23°

c

d

d

8.7 cm

e

44°

10.3 cm

.2
A02
A03

The diagram shows quadrilateral PQRS and its diagonal QS.

Q

8.9 cm

R

108°

5.9 cm

47°

22°

P

S

a In triangle PQS, work out the length of QS.

b In triangle QRS, work out the size of angle RSQ.

c Work out the size of angle PQR.

3
A02

In triangle ABC, BC = 9.2 cm, angle BAC = 56° and angle ABC = 68°.

a Calculate the length of AC.

b Calculate the length of AB.

c Calculate the area of triangle ABC.

4
A02

In triangle XYZ all the angles are acute. XZ = 11.2 cm and XY = 9.4 cm. Angle XYZ = 51°.

a Work out the size of angle XZY.

b Work out the length of YZ.

A☆ 5
A02
A03

The diagram shows the position of three farms, F_1, F_2 and F_3. Farms F_1 and F_2 lie on a straight road such that farm F_2 is 10 km due south of farm F_1. Farm F_3 is on a bearing of 312° from farm F_1. Farm F_2 is on a bearing of 154° from farm F_3. Work out

N

F_3

F_1

10 km

F_2

a the size of angle:

   i $F_1F_3F_2$

   ii $F_1F_2F_3$

b the distance between farm F_1 and farm F_3

c the distance between farm F_2 and farm F_3

d the shortest distance from farm F_1 to the road joining the other two farms.

29.7 The cosine rule

Exercise 29G

Give lengths correct to three significant figures.

A 1 Calculate the length of the sides marked with letters in these triangles.
A02

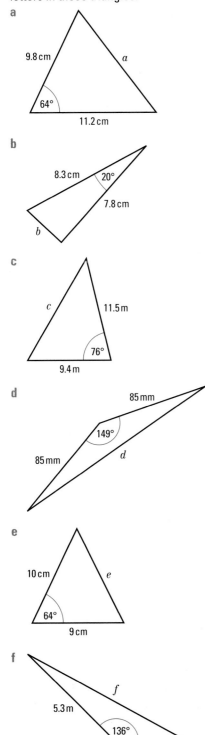

a

9.8 cm

a

64°

11.2 cm

b

8.3 cm

20°

7.8 cm

b

c

c

11.5 m

76°

9.4 m

d

85 mm

149°

d

85 mm

e

10 cm

e

64°

9 cm

f

f

5.3 m

136°

3.1 m

A 2 In triangle ABC, AB = 20.4 cm, AC = 13.5 cm and angle BAC = 42°.
A02
Calculate the length of BC.

29.8 Using the cosine rule to calculate an angle

Exercise 29H

Give lengths and areas correct to three significant figures and angles correct to one decimal place.

A 1 Calculate the size of each of the angles marked with a letter in these triangles.
A02

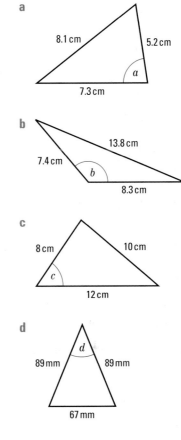

a

8.1 cm

5.2 cm

a

7.3 cm

b

13.8 cm

7.4 cm

b

8.3 cm

c

8 cm

10 cm

c

12 cm

d

d

89 mm

89 mm

67 mm

2 PQ is a chord of a circle with centre O.
A02
The radius of the circle is 6 cm and the length of the chord is 10 cm.

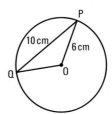

P

10 cm

6 cm

Q

O

Calculate the size of angle POQ.

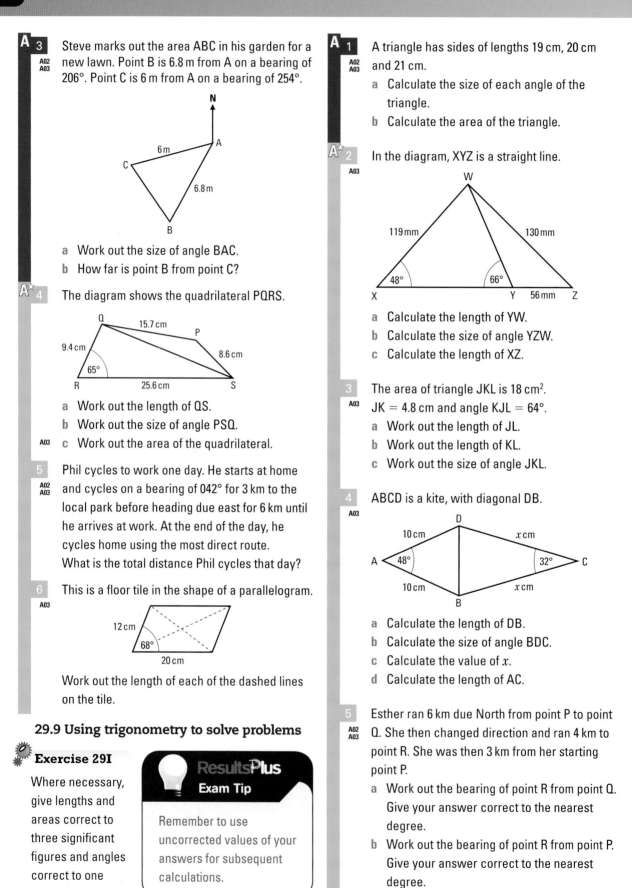

A 3
A02
A03

Steve marks out the area ABC in his garden for a new lawn. Point B is 6.8 m from A on a bearing of 206°. Point C is 6 m from A on a bearing of 254°.

a Work out the size of angle BAC.
b How far is point B from point C?

A★ 4

The diagram shows the quadrilateral PQRS.

a Work out the length of QS.
b Work out the size of angle PSQ.
A03 c Work out the area of the quadrilateral.

5
A02
A03

Phil cycles to work one day. He starts at home and cycles on a bearing of 042° for 3 km to the local park before heading due east for 6 km until he arrives at work. At the end of the day, he cycles home using the most direct route.
What is the total distance Phil cycles that day?

6
A03

This is a floor tile in the shape of a parallelogram.

Work out the length of each of the dashed lines on the tile.

29.9 Using trigonometry to solve problems

Exercise 29I

Where necessary, give lengths and areas correct to three significant figures and angles correct to one decimal place, unless the question states otherwise.

ResultsPlus
Exam Tip

Remember to use uncorrected values of your answers for subsequent calculations.

A 1
A02
A03

A triangle has sides of lengths 19 cm, 20 cm and 21 cm.

a Calculate the size of each angle of the triangle.
b Calculate the area of the triangle.

A★ 2
A03

In the diagram, XYZ is a straight line.

a Calculate the length of YW.
b Calculate the size of angle YZW.
c Calculate the length of XZ.

3
A03

The area of triangle JKL is 18 cm².
JK = 4.8 cm and angle KJL = 64°.

a Work out the length of JL.
b Work out the length of KL.
c Work out the size of angle JKL.

4
A03

ABCD is a kite, with diagonal DB.

a Calculate the length of DB.
b Calculate the size of angle BDC.
c Calculate the value of x.
d Calculate the length of AC.

5
A02
A03

Esther ran 6 km due North from point P to point Q. She then changed direction and ran 4 km to point R. She was then 3 km from her starting point P.

a Work out the bearing of point R from point Q. Give your answer correct to the nearest degree.
b Work out the bearing of point R from point P. Give your answer correct to the nearest degree.

A* **6**

A03

The diagram shows a pyramid. The base of the pyramid, EFGH, is a rectangle in which EF = 10 m and FG = 17 m.
The vertex of the pyramid is P where PE = PF = PG = PH = 22 m.

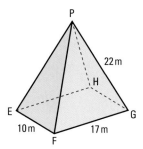

Work out the size of angle FPH.

7

A02
A03

The diagram shows a flag pole, AB, standing on a hill. The hill is at an angle of 9° to the horizontal.
The point C is 18 m downhill from B and the line AC is at 14° to the hill.

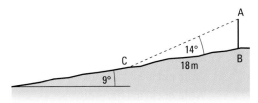

a Calculate the size of angle CAB.
b Calculate the length, AB, of the pole.

A* **8**

A03

A, B and C are points on horizontal ground so that AB = 28 m,
BC = 25 m and
angle CAB = 55°.
AP and BQ are vertical posts, where
AP = BQ = 12 m.

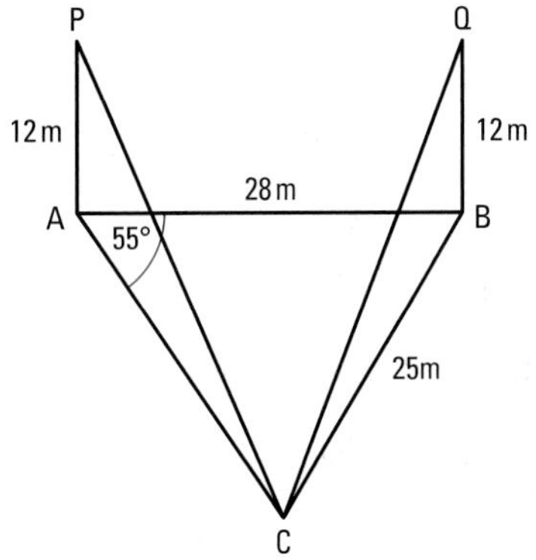

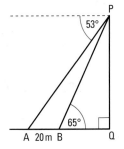

a Work out the size of angle ACB.
b Work out the length of AC.
c Work out the size of angle PCQ.
d Work out the size of the angle between QC and the ground.

9

A02
A03

In the diagram, PQ represents a vertical tower on level ground. The points A and B on the ground are due west of Q. The distance AB is 20 metres. The angle of elevation of P from B is 65°. The angle of depression of A from P is 53°.

Calculate the height of the tower.
Give your answer in metres correct to the nearest metre.

30 Transformations of functions

Key Points

⊙ **function $y = f(x)$:** a rule for working out values of y when given values of x.
⊙ **transformation rules from the curve $y = f(x)$:**
  ⊙ **$y = f(x) + a$:** translation of a units parallel to the y-axis
  ⊙ **$y = f(x + a)$:** translation of $-a$ units parallel to the x-axis
  ⊙ **$y = af(x)$:** stretch of magnitude a parallel to the y-axis
  ⊙ **$y = f(ax)$:** stretch of magnitude $\frac{1}{a}$ parallel to the x-axis
  ⊙ **$y = af\left(\frac{x}{a}\right)$:** enlargement of scale factor a and centre of enlargement (0,0)
  ⊙ **$y = f(-x)$:** reflection in the y-axis
  ⊙ **$y = -f(x)$:** reflection in the x-axis
  ⊙ **$y = -f(-x)$:** rotation by 180° about the origin

30.1 Using function notation

Exercise 30A

> Questions in this chapter are targeted at the grades indicated.

B 1 $f(x) = 4x^3$, $g(x) = \dfrac{5}{2x}$

Find the values of

  a $f(3)$ **b** $f(0)$ **c** $f(-2)$
  d $g(4)$ **e** $g(-2)$ **f** $g\left(\frac{1}{4}\right)$

2 $f(x) = \dfrac{x^2}{2}$, $g(x) = x^2 + 2x + 1$

Find the values of

  a $f(2) + g(2)$ **b** $f(1) + g(4)$

  c $f(3) \times g(3)$ **d** $\dfrac{f(5)}{g(5)}$

A 3 $f(x) = 3x + 1$

  a Find the value of $f(4)$.
  b $f(a) = 28$ Find the value of a.

A☆ 4 $g(x) = 2x^2 - 3$

A02   **a** Find the values of
    i $g(0)$ **ii** $g(1)$ **iii** $g(-4)$.
  b $g(k) = 15$ Find the values of k.

5 $g(x) = x(x - 2)$

A02   **a** Find the values of
    i $g(3)$ **ii** $g(-2)$ **iii** $4g(7)$.
  b $g(n) = 0$ Find the values of n.

A☆ 6 $f(x) = 2x(x - 1)$

A02
A03   **a** Find the values of
    i $f(1)$ **ii** $f(4)$ **iii** $2f(-3)$.
  b $f(h) = 0$ Find the values of h.
  c $f(m) = 12$ Find the values of m.

7 $f(x) = 2x^2$

A02   **a** Find $f(3)$.
  b Write out in full $f(x) + 6$.
  c Write out in full $f(x - 8)$.

8 $g(x) = 3(x + 2)$

A02   **a** Find $g(-4)$.
  b Write out in full $g(-5x)$.
  c Write out in full $6g(x)$.

30.2 Translation of a curve parallel to the axes

Exercise 30B

A☆ 1 Here is the graph of $y = f(x) = x^2$.

A03

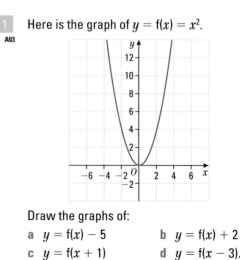

Draw the graphs of:

  a $y = f(x) - 5$ **b** $y = f(x) + 2$
  c $y = f(x + 1)$ **d** $y = f(x - 3)$.

2 Here is a sketch of the graph of $y = f(x) = x^3$.

a Draw sketches of the graphs of:
  i $y = f(x) - 1$
  ii $y = f(x + 3)$.

b Write down the coordinates of the point to which the point (0, 0) is mapped in each case.

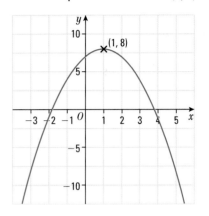

3 Here is a sketch of the graph of $y = f(x) = \dfrac{3}{x}$.

The curve $y = f(x)$ is translated by $\begin{pmatrix} 0 \\ -3 \end{pmatrix}$.

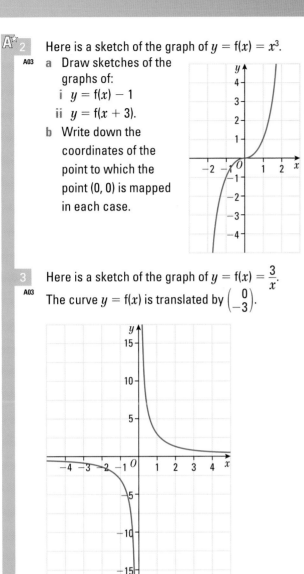

a Sketch the graph of the new curve.

b Write down the coordinates to which the point (2, 1.5) is mapped.

c Write down the equation of the translated curve:
  i in function form
  ii in algebraic form.

The curve $y = f(x)$ is now translated by $\begin{pmatrix} -3 \\ 0 \end{pmatrix}$

d Sketch the transformed curve.

e The point (1, 6) is mapped to the point (p, q). Write down the values of p and q.

f Write down the equation of the translated curve:
  i in function form
  ii in algebraic form.

4 Here is a curve with equation $y = f(x)$.
The maximum point of the curve is (1, 8).

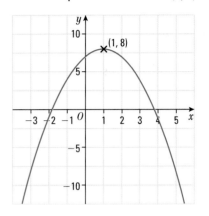

a Write down the coordinates of the maximum point of $y = f(x) + 2$.

b Write down the coordinates of the maximum point of $y = f(x + 3)$.

5 Here are two curves, C_1 and C_2. The equation of the curve C_1 is $y = f(x)$.
The curve C_1 can be mapped to the curve C_2 by a translation.
The minimum point of C_1 is (1, −4) and the minimum point of C_2 is (1, −9).

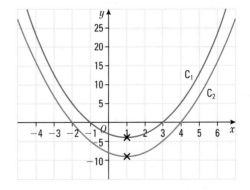

a Describe the translation.

b Write down the equation of the curve C_2 in function form.

The algebraic equation of the curve C_1 is $y = x^2 - 2x - 3$.

c Write down the algebraic equation of the curve C_2.

6 Here are two curves, C_1 and C_2. The equation of the curve C_1 is $y = f(x)$.

The curve C_1 can be mapped to the curve C_2 by a translation.

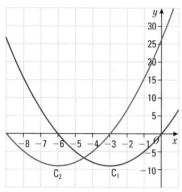

a Describe the translation.

b Write down the equation of the curve C_2 in function form.

The algebraic equation of the curve C_1 is $y = x^2 + 6x$.

c Write down the algebraic equation of the curve C_2.

7 The expression $x^2 + 6x + 11$ can be written in the form $(x + a)^2 + b$ for all values of x.

a Find the value of a and the value of b.

The graph of $x^2 + 6x + 11$ can be obtained from the graph of $y = x^2$ by a translation.

b Describe this translation.

c Sketch the graph of $y = x^2$.

d Sketch the graph of $x^2 + 6x + 11$ on the same axes.

8 Describe fully the transformation that will map the curve with equation $y = x^2$ to the curve with equation $y = x^2 + 8$.

30.3 Stretching a curve parallel to the axes

Exercise 30C

1 Here is the graph of $y = f(x)$. It has a minimum point at $(2, 2)$.
Copy the graph and on the same grid draw the graph of $y = 3f(x)$.
To which point is the minimum point of $y = f(x)$ mapped?

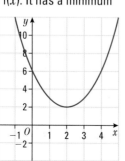

2 Here is a sketch of the curve C_1 $y = f(x) = x^2$.

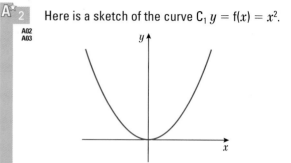

a Copy the sketch and on the same axes sketch the curve C_2 with equation $y = f(4x)$.

b Write down the equation of the curve C_2 in algebraic form.

c Give two different transformations that will each map the curve C_1 to the curve C_2.

3 Here is the curve C_1 with equation $y = f(x) = \sin x°$.

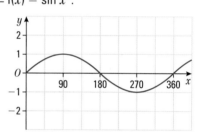

a Draw the curve C_2 with equation $y = f(2x)$.

b Describe the transformation that maps C_1 onto C_2.

c Write down the algebraic form of the equation $y = f(2x)$.

4 Here is the graph of $y = f(x)$.
The graph crosses the y-axis at $(0, 8)$ and the x-axis at $(-2, 0)$.

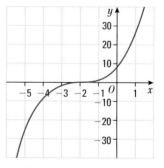

a Sketch the graph with equation $y = 2f(x)$.

b Write down the coordinates of the points to which $(0, 8)$ and $(-2, 0)$ are mapped.

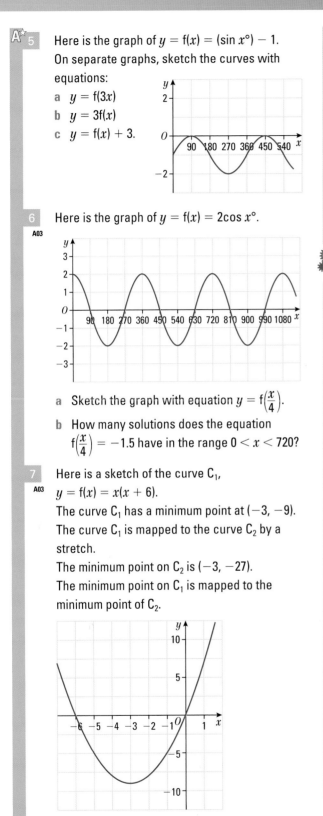

A★ 5 Here is the graph of $y = f(x) = (\sin x°) - 1$.
On separate graphs, sketch the curves with equations:

a $y = f(3x)$

b $y = 3f(x)$

c $y = f(x) + 3$.

6 Here is the graph of $y = f(x) = 2\cos x°$.
A03

a Sketch the graph with equation $y = f\left(\dfrac{x}{4}\right)$.

b How many solutions does the equation $f\left(\dfrac{x}{4}\right) = -1.5$ have in the range $0 < x < 720$?

7 Here is a sketch of the curve C_1,
A03 $y = f(x) = x(x + 6)$.
The curve C_1 has a minimum point at $(-3, -9)$.
The curve C_1 is mapped to the curve C_2 by a stretch.
The minimum point on C_2 is $(-3, -27)$.
The minimum point on C_1 is mapped to the minimum point of C_2.

a Describe the stretch fully.

b Draw a sketch of C_2.

c Write the equation of C_2:

  i using functional form

  ii in algebraic form.

A★ 8 The expression $x^2 - 10x + 7$ can be written in
A02 the form $(x - p)^2 + q$.
A03

a Find the values of p and q.

b Write down the coordinates of the point P where the curve $y = f(x) = x^2 - 10x + 7$ crosses the y-axis.

The curve $y = f(x)$ is mapped by a stretch parallel to the y-axis, so that the point P is mapped to the point $(0, 4)$.

c Describe the stretch and write down the equation of the new curve.

Exercise 30D

A★ 1 Here is a graph of the curve with equation
A03 $y = \sin x°$.

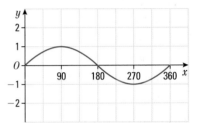

a Copy the graph and sketch the curve C_2, the enlargement of C_1 with scale factor 3.

b Write down the equation of the curve C_2.

2 a Draw a sketch of the U-shaped curve C_1, with
A03 equation $y = (x - 2)^2$.

The curve C_1 is enlarged with a scale factor 2, centre O, to give the curve C_2.

b Find the equation of the curve C_2 in function form.

c Sketch the curve C_2.

3 a Sketch the curve with equation $y = f(x) = $
A03 $\cos x°$ for values of x from 0 to 1080.

b On the same axes, sketch the graph of the curve with equation $y = 2f\left(\dfrac{x}{2}\right)$.

4 Write down the algebriac equation of the given
A03 curve after it has been enlarged by scale factor 2 and centre O.

a $y = x^2 + 2$ b $y = \dfrac{1}{x} + 2$

c $y = 2^x$ d $y = 3\sin(2x)$

30.4 Rotation about the origin and reflection in the axes

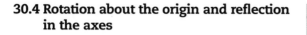

Exercise 30E

A★ 1
A03

The graph of $y = f(x) = 1 - x^2$ has been drawn.

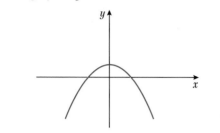

a Sketch the graph of $y = -f(x)$.
b Write down the equation of the new graph in algebraic form.

2
A03

Here is the graph of $y = f(x)$.

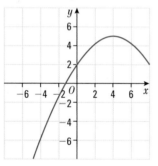

Sketch:
a $y = f(-x)$
b $y = f(2x)$
c Hence, or otherwise, sketch $y = f(-2x)$.

A★ 3
A03

Here is the graph of the curve C_1 $y = f(x)$.
C_2 is the image of C_1 under a rotation of 180° about the origin.

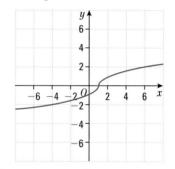

a Sketch the curve C_2.
b Write down the equation of the curve C_2 in function form.

31 Circle geometry

Key Points

⊙ **circle facts:**

- ⊙ a triangle in a circle where two of its sides are radii is an isosceles triangle

- ⊙ the angle between a tangent and a radius of a circle is 90°

- ⊙ tangents to a circle from a point outside the circle are equal in length

⊙ **circle theorems:**

- ⊙ **1)** the perpendicular from the centre of a circle to a chord bisects the chord (and a line drawn from the centre of a circle to the midpoint of a chord is perpendicular to the chord)

- ⊙ **2)** the angle at the centre of a circle is twice the angle at the circumference, where both angles are subtended by the same arc

- ⊙ **3)** the angle in a semicircle is a right angle

- ⊙ **4)** angles in the same segment are equal

- ⊙ **5)** opposite angles of a cyclic quadrilateral add up to 180°

- ⊙ **6)** the angle between a tangent and a chord is equal to the angle in the alternate segment

31.1 Isosceles triangle in a circle

Exercise 31A

> Questions in this chapter are targeted at the grades indicated.

The diagrams all show circles, centre O.
Work out the size of each angle marked with a letter.

D 1

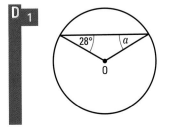

D 2

3

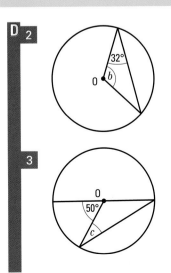

In questions 4–6 give reasons for your answers.

Results Plus
Exam Tip

The reasons you give should be the rules that you have used and should be written in words.

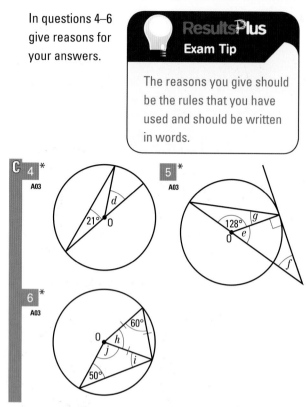

C 4 *
A03

5 *
A03

6 *
A03

In questions 4–6 give reasons for your answers.

B 4
A03

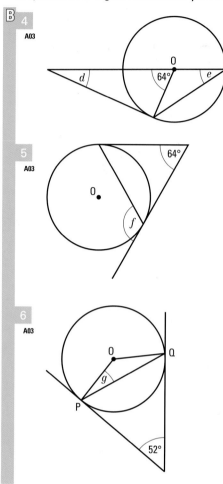

5
A03

6
A03

31.2 Tangents to a circle

Exercise 31B

The diagrams all show circles, centre O.
Work out the size of each angle marked with a letter.

B 1

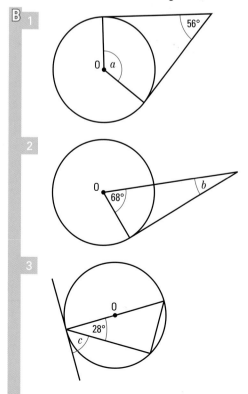

2

3

31.3 Circle theorems

Exercise 31C

Find the size of each of the angles marked with a letter.
O is the centre of the circle in each case.

A 1

2

3

4

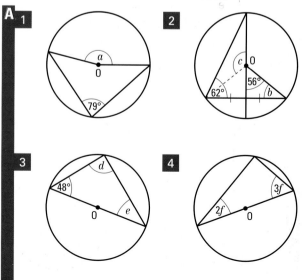

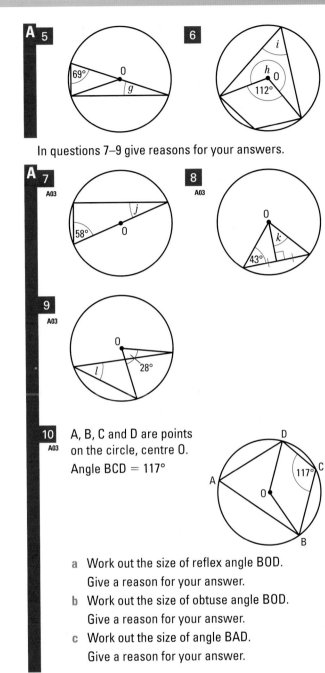

A **5**

69° 0
g

6

i
h 0
112°

In questions 7–9 give reasons for your answers.

A **7**
A03

58°
0
j

8
A03

0
k
43°

9
A03

0
l
28°

A **3**
A03

c
66°
•0

4
A03

55° 87°
0
d

5
A03

52°
0 e
g
f

6
A03

0
40°
h

In questions 7–9 give reasons for your answers.

A **7** *
A03

D
70°
i
A
34°
•0
B
C

8 *
A03

E
j
H 106°
•0
F
G

9 *
A03

I
0
56°
k
H 52°
J

10
A03

A, B, C and D are points on the circle, centre O.
Angle BCD = 117°

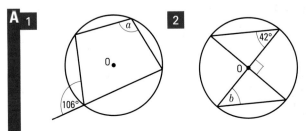

D
A
0
117° C
B

a Work out the size of reflex angle BOD.
  Give a reason for your answer.
b Work out the size of obtuse angle BOD.
  Give a reason for your answer.
c Work out the size of angle BAD.
  Give a reason for your answer.

31.4 More circle theorems

Exercise 31D

Find the size of each of the angles marked with a letter.
O is the centre of the circle in each case.

A **1**

a
0.
106°

2

42°
0
b

32 Algebraic fractions and algebraic proof

Key Points

- **simplifying algebraic fractions:** factorise the numerator and the denominator, then divide the numerator and denominator by any common factors.
- **adding and subtracting algebraic fractions:**
  - if the denominators are the same, add or subtract the numerators but do not change the denominator
  - if the denominators are different, write each fraction as an equivalent fraction with a common denominator
- **multiplying algebraic fractions:** multiply the numerators and multiply the denominators.
- **dividing algebraic fractions:** multiply the first fraction by the reciprocal of the second.

- **proving a result is true:** you must show that it will be true in all cases. The following points may be helpful:
  - consecutive integers can be written in the form $n, n + 1, n + 2, \ldots$
  - any even number can be written in the form $2n$
  - consecutive even numbers can be written in the form $2n, 2n + 2, 2n + 4, \ldots$
  - any odd number can be written in the form $2n - 1$ (or, e.g. $2n + 1$)
  - consecutive odd numbers can be written in the form $2n - 1, 2n + 1, 2n + 3, \ldots$

32.1 Simplifying algebraic fractions

Exercise 32A

Questions in this chapter are targeted at the grades indicated.

A **1** Simplify fully.

a $\dfrac{5x^6}{x^2}$ b $\dfrac{3x^3y}{6xy^2}$ c $\dfrac{x^3 - 4x}{2x}$

d $\dfrac{x^2 - x}{x - 1}$ e $\dfrac{4x - 8x^2}{2x - 1}$

A **2** Simplify fully.

a $\dfrac{x^2 + 7x + 12}{x^2 + 5x + 6}$ b $\dfrac{x^2 + 7x + 10}{x^2 + 2x}$

c $\dfrac{x^2 + x - 6}{x^2 + 6x + 9}$ d $\dfrac{x^2 - 8x + 15}{x^2 + x - 12}$

3 Simplify fully.

a $\dfrac{x^2 - 1}{x^2 + x}$ b $\dfrac{3x^2 - 12}{x^2 - 4x + 4}$

c $\dfrac{2x^2 - 18x}{x^2 - 81}$ d $\dfrac{2x^2 - 18}{3x^2 - 9x}$

4 Simplify fully.

a $\dfrac{9x^2 - 1}{9x^2 - 9x + 2}$ b $\dfrac{2x^2 + 5x + 3}{3x^2 + x - 2}$

c $\dfrac{14x^2 - 19x - 3}{8x^2 - 14x + 3}$ d $\dfrac{12x^2 + 4x - 1}{12x^2 + 12x + 3}$

A **5** Simplify fully.

a $\dfrac{x(x + 3)}{x^2 - 3x}$ b $\dfrac{8x^2 - 10x + 3}{8x - 6}$

c $\dfrac{x^2 + 8x + 15}{3x^2 - 27}$ d $\dfrac{9 - x^2}{(x + 3)^2}$

e $\dfrac{2x^3 + 4x^2}{6x^2 - 4x^3}$ f $\dfrac{25 - x^2}{5 - x}$

32.2 Adding and subtracting algebraic fractions

Exercise 32B

A **1** Write as a single fraction in its simplest form.

a $\dfrac{2x}{5} + \dfrac{2x}{5}$ b $\dfrac{x}{4} + \dfrac{3x}{4}$ c $\dfrac{1}{12x} + \dfrac{3}{12x}$

d $\dfrac{7x}{8} - \dfrac{3x}{8}$ e $\dfrac{5}{3x} - \dfrac{1}{3x}$ f $\dfrac{4x}{7} - \dfrac{3x}{7}$

2 Write as a single fraction in its simplest form.

a $\dfrac{x}{3} + \dfrac{x}{5}$ b $\dfrac{x}{3} + \dfrac{2x}{15}$ c $\dfrac{3x}{4} - \dfrac{x}{8}$

d $\dfrac{3x}{4} - \dfrac{x}{3}$ e $\dfrac{1}{3x} + \dfrac{1}{4x}$ f $\dfrac{4}{15x} - \dfrac{3}{30x}$

A **3** Simplify.

a $\dfrac{x}{2} + \dfrac{x - 1}{9}$ b $\dfrac{x - 3}{5} + \dfrac{x + 2}{4}$

c $\dfrac{2x}{5} - \dfrac{5x}{7}$ d $\dfrac{1}{x + 1} + \dfrac{1}{x + 2}$

e $\dfrac{3}{x + 2} - \dfrac{4}{x + 1}$ f $\dfrac{1}{x - 1} - \dfrac{1}{2x + 1}$

Exercise 32C

D 1 **a** Factorise **i** $4x - 4$ **ii** $24x - 24$.
   b Write down the lowest common multiple of $4x - 4$ and $24x - 24$.
   c Write $\dfrac{1}{4x - 4} - \dfrac{3}{24x - 24}$ as a single fraction in its simplest form.

B 2 Write down the lowest common multiple of each of the following pairs of expressions.
   a $3x$ and $4x$ **b** $x + 2$ and $x + 1$
   c $2x$ and $x(x - 2)$ **d** $x - 3$ and $(x + 4)(x - 3)$
   e $x^2 + 2x$ and $x + 2$
   f $3x + 12$ and $x + 4$

A★ 3 **a** Factorise $x^2 - 6x + 9$.
   b Write $\dfrac{1}{x - 3} - \dfrac{1}{x^2 - 6x + 9}$ as a single fraction in its simplest form.

4 **a** Factorise $x^2 - 9$.
   b Write $\dfrac{3}{x - 3} - \dfrac{2}{x^2 - 9}$ as a single fraction in its simplest form.

5 **a** Factorise $2x^2 - x - 1$.
   b Write $\dfrac{1}{2x^2 - x - 1} + \dfrac{2}{2x + 1}$ as a single fraction in its simplest form.

6 Simplify $\dfrac{1}{2x + 10} - \dfrac{1}{x^2 + 6x + 5}$

7 Write $\dfrac{1}{6} + \dfrac{1}{2x} + \dfrac{1}{3(x + 1)}$ as a single fraction.

8 Express $\dfrac{1}{x + 1} - \dfrac{2}{x^2 - 1}$ as a single fraction.

9 **a** Factorise **i** $x^2 - 4$ **ii** $x^2 + 2x$.
   b Write $\dfrac{1}{x^2 - 4} + \dfrac{1}{x^2 + 2x}$ as a single fraction in its simplest form.

10 Show that $\dfrac{1}{9x^2 - 3x - 2} - \dfrac{1}{9x^2 - 1}$
   $= \dfrac{A}{(3x - 1)(3x + 1)(3x - 2)}$ and find the value of A.

32.3 Multiplying and dividing algebraic fractions

Exercise 32D

C 1 Write as a single fraction.
   a $\dfrac{x}{3} \times \dfrac{x}{4}$ **b** $\dfrac{6}{y} \times \dfrac{2}{y}$
   c $\dfrac{3x}{2} \times \dfrac{9y}{5}$ **d** $\dfrac{x}{4} \times \dfrac{x + 2}{5}$

B 2 Write as a single fraction in its simplest form.
   a $\dfrac{3x}{2} \times \dfrac{7y}{6}$ **b** $\dfrac{6y}{7x} \times \dfrac{x}{2}$
   c $\dfrac{3x^3}{y^2} \times \dfrac{y}{x}$ **d** $\dfrac{x + 2}{x} \times \dfrac{2x}{x - 2}$

3 Write as a single fraction.
   a $\dfrac{x}{7} \div \dfrac{x}{5}$ **b** $\dfrac{5x}{8} \div \dfrac{2}{y}$
   c $x^3y \div \dfrac{1}{y^2}$ **d** $\dfrac{2x}{x + 3} \div \dfrac{x + 3}{x + 2}$

4 Write as a single fraction in its simplest form.
   a $\dfrac{5x}{4} \div \dfrac{4x}{5}$ **b** $\dfrac{4}{y} \div \dfrac{12y}{7x}$
   c $\dfrac{3y^2}{8x} \div \dfrac{y}{x^2}$ **d** $\dfrac{6}{y + 3} \div \dfrac{2}{y}$

A 5 Write as a single fraction in its simplest form.
   a $\dfrac{x + 1}{2} \times \dfrac{3x + 3}{4}$ **b** $\dfrac{x + 2}{x - 2} \times (x - 2)^2$
   c $\dfrac{x}{x + 3} \times \dfrac{x + 3}{x + 2}$ **d** $\dfrac{2x + 12}{x - 5} \div \dfrac{x + 6}{3}$
   e $\dfrac{6}{5x + 2} \div \dfrac{2}{(5x + 2)^2}$ **f** $\dfrac{x + 5}{12} \div \dfrac{2x + 10}{4}$

A★ 6 **a** Factorise $x^2 - 16$.
   b Write $\dfrac{1}{x + 4} \times \dfrac{x^2 - 16}{x^2 + 16}$ as a single fraction in its simplest form.

7 **a** Factorise **i** $x^2 + 7x + 12$ **ii** $x^2 + 10x + 24$.
   b Write $\dfrac{x + 6}{x^2 + 7x + 12} \div \dfrac{x + 3}{x^2 + 10x + 24}$ as a single fraction in its simplest form.

8 Write $\dfrac{x^2 - 2x}{x^2 + 2x} \times \dfrac{x^2 + 4x + 4}{x - 2}$ as a single fraction in its simplest form.

32.4 Algebraic proof

Exercise 32E

A★ 1 Prove that the difference between any odd number and any even number is odd. [A03]

2 * Prove that the square of any odd number is odd. [A03]

3 * Prove that the sum of any five consecutive numbers is a multiple of 5. [A03]

4 * **a** Prove that the sum of any two consecutive multiples of 5 is odd. [A03]
   b Prove that the product of any two consecutive multiples of 5 is even.

5 * Prove that for any two numbers n_1 and n_2, $(n_1^2 - n_2^2) - (n_1 - n_2)^2$ is even. [A03]

6 * Prove that, if the difference of two numbers is 3, then the difference of their squares is a multiple of 3. [A03]

33 Vectors

Key Points

⊙ **vector:** a quantity that has magnitude and direction.
- ⊙ labelled with a single bold letter (e.g. **a**, **b**, **c**) or they can be handwritten using underlined letters (e.g. a̲, b̲, c̲)
- ⊙ equal vectors must have the same magnitude and direction

⊙ **displacement:** a change in position.
- ⊙ a displacement from A to B can be written as $\overrightarrow{AB}$ to show that it is a vector
- ⊙ can be described as a column vector, e.g. $\begin{pmatrix} 4 \\ 2 \end{pmatrix}$ which means 4 units to the right and 2 units up

⊙ **magnitude:** the size of the vector.
- ⊙ the magnitude of the vector **a** is written a or $|a|$
- ⊙ the magnitude of the vector $\overrightarrow{AB}$ is the length of the line segment AB
- ⊙ the magnitude of the vector $\begin{pmatrix} x \\ y \end{pmatrix}$ is $\sqrt{x^2 + y^2}$

⊙ **triangle law of vector addition:**
$$\overrightarrow{AB} + \overrightarrow{BC} = \overrightarrow{AC}$$
$$\mathbf{a} + \mathbf{b} = \mathbf{c}$$

⊙ **parallelogram law of vector addition:**
$$\overrightarrow{PQ} = \overrightarrow{SR} = \mathbf{a}$$
$$\overrightarrow{PS} = \overrightarrow{QR} = \mathbf{b}$$
$$\overrightarrow{PR} = \overrightarrow{PQ} + \overrightarrow{QR} = \mathbf{a} + \mathbf{b}$$
$$\overrightarrow{PR} = \overrightarrow{PQ} + \overrightarrow{PS}$$

⊙ **resultant vector:** the vector **c** if **c** = **a** + **b**.

⊙ **zero displacement:** occurs when there is no overall change in position, e.g. $\overrightarrow{AB} + \overrightarrow{BA} = 0$.

⊙ **scalar:** a non-zero magnitude, k, that is applied to a vector.
- ⊙ if $\mathbf{a} = \begin{pmatrix} p \\ q \end{pmatrix}$, then $k\mathbf{a} = k\begin{pmatrix} p \\ q \end{pmatrix} = \begin{pmatrix} kp \\ kq \end{pmatrix}$;
- ⊙ vectors **a** and $k\mathbf{a}$ are parallel
- ⊙ $-\overrightarrow{BA} = \overrightarrow{AB}$

⊙ **position vector:** a vector which joins the origin to a point. $\overrightarrow{OA}$ is the position vector of point A.
- ⊙ point (p, q) has position vector $\begin{pmatrix} p \\ q \end{pmatrix}$

⊙ **adding column vectors:** use the rule
$$\begin{pmatrix} a \\ b \end{pmatrix} + \begin{pmatrix} c \\ d \end{pmatrix} = \begin{pmatrix} a + c \\ b + d \end{pmatrix}$$

33.1 Vectors and vector notation

Exercise 33A

Questions in this chapter are targeted at the grades indicated.

A 1 On squared paper draw and label the following vectors.

a $\mathbf{a} = \begin{pmatrix} 2 \\ 3 \end{pmatrix}$ b $\mathbf{b} = \begin{pmatrix} 4 \\ -1 \end{pmatrix}$ c $\mathbf{c} = \begin{pmatrix} -5 \\ -2 \end{pmatrix}$

d $\overrightarrow{AB} = \begin{pmatrix} -4 \\ 5 \end{pmatrix}$ e $\overrightarrow{CD} = \begin{pmatrix} 0 \\ 3 \end{pmatrix}$

2 The point X is (1, 2), the point Y is (5, 9) and the point Z is (5, −1).

a Write as column vectors:

  i $\overrightarrow{XY}$ ii $\overrightarrow{YZ}$ iii $\overrightarrow{XZ}$

b What do you notice about your answers in a?

A 3 The points A, B, C and D are the vertices of a quadrilateral where A has coordinates (2, 2),

$\overrightarrow{AB} = \begin{pmatrix} 0 \\ 3 \end{pmatrix}$, $\overrightarrow{BC} = \begin{pmatrix} 3 \\ 0 \end{pmatrix}$ and $\overrightarrow{CD} = \begin{pmatrix} 2 \\ -5 \end{pmatrix}$.

a On squared paper draw quadrilateral ABCD.
b Write as a column vector $\overrightarrow{AD}$.
c What type of quadrilateral is ABCD?
d What do you notice about $\overrightarrow{AB}$ and $\overrightarrow{BC}$?

4 The points A, B, C and D are the vertices of a rhombus.

A has coordinates (4, 2), $\overrightarrow{AB} = \begin{pmatrix} -3 \\ 2 \end{pmatrix}$ and $\overrightarrow{AD} = \begin{pmatrix} 3 \\ 2 \end{pmatrix}$.

a On squared paper draw the rhombus ABCD.
b Write as a column vector i $\overrightarrow{CD}$ ii $\overrightarrow{BC}$
c What do you notice about
  i $\overrightarrow{AB}$ and $\overrightarrow{CD}$ ii $\overrightarrow{AD}$ and $\overrightarrow{BC}$?

A 5 Here are eight vectors.

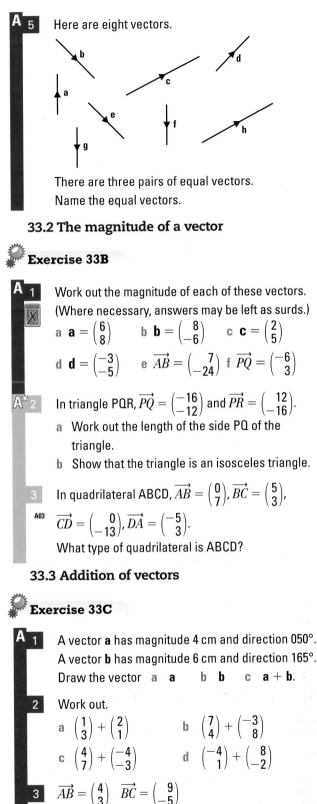

There are three pairs of equal vectors.
Name the equal vectors.

33.2 The magnitude of a vector

Exercise 33B

A 1 Work out the magnitude of each of these vectors.
(Where necessary, answers may be left as surds.)

a $\mathbf{a} = \begin{pmatrix} 6 \\ 8 \end{pmatrix}$ b $\mathbf{b} = \begin{pmatrix} 8 \\ -6 \end{pmatrix}$ c $\mathbf{c} = \begin{pmatrix} 2 \\ 5 \end{pmatrix}$

d $\mathbf{d} = \begin{pmatrix} -3 \\ -5 \end{pmatrix}$ e $\overrightarrow{AB} = \begin{pmatrix} 7 \\ -24 \end{pmatrix}$ f $\overrightarrow{PQ} = \begin{pmatrix} -6 \\ 3 \end{pmatrix}$

A* 2 In triangle PQR, $\overrightarrow{PQ} = \begin{pmatrix} -16 \\ -12 \end{pmatrix}$ and $\overrightarrow{PR} = \begin{pmatrix} 12 \\ -16 \end{pmatrix}$.

a Work out the length of the side PQ of the triangle.

b Show that the triangle is an isosceles triangle.

3 In quadrilateral ABCD, $\overrightarrow{AB} = \begin{pmatrix} 0 \\ 7 \end{pmatrix}$, $\overrightarrow{BC} = \begin{pmatrix} 5 \\ 3 \end{pmatrix}$,

A03 $\overrightarrow{CD} = \begin{pmatrix} 0 \\ -13 \end{pmatrix}$, $\overrightarrow{DA} = \begin{pmatrix} -5 \\ 3 \end{pmatrix}$.

What type of quadrilateral is ABCD?

33.3 Addition of vectors

Exercise 33C

A 1 A vector **a** has magnitude 4 cm and direction 050°.
A vector **b** has magnitude 6 cm and direction 165°.
Draw the vector   a **a**   b **b**   c **a + b**.

2 Work out.

a $\begin{pmatrix} 1 \\ 3 \end{pmatrix} + \begin{pmatrix} 2 \\ 1 \end{pmatrix}$ b $\begin{pmatrix} 7 \\ 4 \end{pmatrix} + \begin{pmatrix} -3 \\ 8 \end{pmatrix}$

c $\begin{pmatrix} 4 \\ 7 \end{pmatrix} + \begin{pmatrix} -4 \\ -3 \end{pmatrix}$ d $\begin{pmatrix} -4 \\ 1 \end{pmatrix} + \begin{pmatrix} 8 \\ -2 \end{pmatrix}$

3 $\overrightarrow{AB} = \begin{pmatrix} 4 \\ 3 \end{pmatrix}$ $\overrightarrow{BC} = \begin{pmatrix} 9 \\ -5 \end{pmatrix}$

Work out $\overrightarrow{AC}$.

4 $\mathbf{s} = \begin{pmatrix} 2 \\ 4 \end{pmatrix}$ $\mathbf{t} = \begin{pmatrix} 3 \\ 5 \end{pmatrix}$ $\mathbf{u} = \begin{pmatrix} 1 \\ -6 \end{pmatrix}$

a Work out i **s + t + u**   ii **u + t + s**

b What do you notice?

A* 5 ABCDEFGH is a regular octagon.
A02
A03 $\overrightarrow{AB} = \mathbf{a}$ and $\overrightarrow{BC} = \mathbf{b}$.

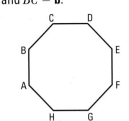

a Which other displacement is equal to the vector **a**?

b Which other displacement is equal to the vector **b**?
$\overrightarrow{CD} = \mathbf{c}, \overrightarrow{DE} = \mathbf{d}$

c Find:
   i $\overrightarrow{AE}$ ii $\overrightarrow{GE}$ iii $\overrightarrow{HE}$.

33.4 Parallel vectors

Exercise 33D

A 1 The vector **a** has magnitude 5 cm and direction 080°.
The vector **b** has magnitude 6 cm and direction 170°.
Draw the vector

a **a**   b **b**   c **−a**   d **b − a**.

2 Here is the vector **q**.
Draw the vector   a **3q**   b **−⅓q**.

A* 3 $\mathbf{m} = \begin{pmatrix} 5 \\ 4 \end{pmatrix}$ $\mathbf{n} = \begin{pmatrix} -3 \\ 7 \end{pmatrix}$ $\mathbf{p} = \begin{pmatrix} 7 \\ -4 \end{pmatrix}$

a Find as a column vector.
   i **7m**   ii **−3n**   iii **4m + 2p**
   iv **2m + 3n − 4p**

b Find   i the magnitude of the vector **m**
   ii the magnitude of the vector **m − 3p**.

4 The points P, Q, R and S have coordinates
A02
A03 $(-2, 4), (4, 2), (-6, -8)$ and $(12, -14)$ respectively.

a Write down the position vector,
   $\overrightarrow{OP}$, of the point P.

b Write down as a column vector.
   i $\overrightarrow{PQ}$ ii $\overrightarrow{RS}$.

c What do these results show about the lines PQ and RS?

5 The point A has coordinates (2, 5), the point
A02
A03 B has coordinates (5, 6), the point C has coordinates $(-1, -2)$.
Find the coordinates of the point D where $\overrightarrow{CD} = 5\overrightarrow{AB}$.

A* 6 $\overrightarrow{OP} = \mathbf{p}$

$\overrightarrow{OQ} = \mathbf{q}$

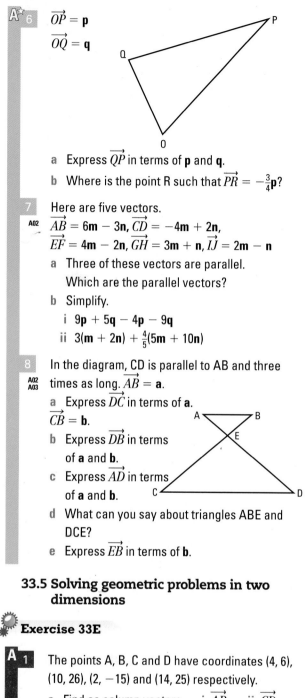

a Express $\overrightarrow{QP}$ in terms of $\mathbf{p}$ and $\mathbf{q}$.

b Where is the point R such that $\overrightarrow{PR} = -\frac{3}{4}\mathbf{p}$?

7 Here are five vectors.

A02 $\overrightarrow{AB} = 6\mathbf{m} - 3\mathbf{n}, \overrightarrow{CD} = -4\mathbf{m} + 2\mathbf{n},$
$\overrightarrow{EF} = 4\mathbf{m} - 2\mathbf{n}, \overrightarrow{GH} = 3\mathbf{m} + \mathbf{n}, \overrightarrow{IJ} = 2\mathbf{m} - \mathbf{n}$

a Three of these vectors are parallel.
   Which are the parallel vectors?

b Simplify.

  i $9\mathbf{p} + 5\mathbf{q} - 4\mathbf{p} - 9\mathbf{q}$

  ii $3(\mathbf{m} + 2\mathbf{n}) + \frac{4}{5}(5\mathbf{m} + 10\mathbf{n})$

8 In the diagram, CD is parallel to AB and three
A02 times as long. $\overrightarrow{AB} = \mathbf{a}$.
A03
a Express $\overrightarrow{DC}$ in terms of $\mathbf{a}$.
$\overrightarrow{CB} = \mathbf{b}$.

b Express $\overrightarrow{DB}$ in terms
   of $\mathbf{a}$ and $\mathbf{b}$.

c Express $\overrightarrow{AD}$ in terms
   of $\mathbf{a}$ and $\mathbf{b}$.

d What can you say about triangles ABE and
   DCE?

e Express $\overrightarrow{EB}$ in terms of $\mathbf{b}$.

33.5 Solving geometric problems in two dimensions

Exercise 33E

A 1 The points A, B, C and D have coordinates (4, 6),
(10, 26), (2, −15) and (14, 25) respectively.

a Find as column vectors i $\overrightarrow{AB}$ ii $\overrightarrow{CD}$.

b What do these results show about the line
   segments AB and CD?

A* 2 In triangle OPQ, $\overrightarrow{OP} = \mathbf{p}$ and $\overrightarrow{OQ} = \mathbf{q}$.
A02
A03

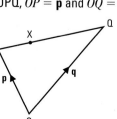

A* a Find in terms of $\mathbf{p}$ and $\mathbf{q}$ the vector $\overrightarrow{QP}$.
X is a point that lies a third of the way along PQ.

b Find in terms of $\mathbf{p}$ and $\mathbf{q}$ the vector $\overrightarrow{QX}$.

c Find in terms of $\mathbf{p}$ and $\mathbf{q}$ the vector $\overrightarrow{OX}$.

3 OPRQ is a parallelogram with
A02 $\overrightarrow{OP} = \mathbf{p}$ and $\overrightarrow{OQ} = \mathbf{q}$.
A03
X is a point that lies a
third of the way along PQ.

a Use the result of
   question 2 to write down $\overrightarrow{QX}$ in terms of
   $\mathbf{p}$ and $\mathbf{q}$.

b Express $\overrightarrow{OR}$ in terms of $\mathbf{p}$ and $\mathbf{q}$.
Y is a point that lies two thirds of the way
along OR.

c Express $\overrightarrow{QY}$ in terms of $\mathbf{p}$ and $\mathbf{q}$.

d Express $\overrightarrow{XY}$ in terms of $\mathbf{p}$ and $\mathbf{q}$.

e What does your answer to d tell you about
   PR, XY and OQ?

4 ABCD is a quadrilateral. A is the point (1, 2), B is
A02 the point (3, 6) and P is the midpoint of AB.
A03
a Write down the coordinates of P.
C is the point (7, 4) and Q is the midpoint of BC.

b Write down the coordinates of Q.
D is the point (7, 2) and R is the midpoint of CD.

c Write down the coordinates of R.
S is the midpoint of AD.

d Write down the coordinates of S.

e Find as column vectors: i $\overrightarrow{PQ}$ ii $\overrightarrow{SR}$.

f Explain with reasons what the answers to e
   show about the quadrilateral PQRS.

5 OXYZ is a quadrilateral.
A02 $\overrightarrow{OX} = \mathbf{a}, \overrightarrow{OY} = \mathbf{a} + \mathbf{b}$ and $\overrightarrow{ZY} = \mathbf{a} - \mathbf{b}$.
A03

a Find in terms of $\mathbf{a}$ and $\mathbf{b}$: i $\overrightarrow{OZ}$ ii $\overrightarrow{XY}$.

b What do your answers to a show about OZ
   and XY?
The point P is such that $\overrightarrow{YP} = 2\mathbf{a} - \mathbf{b}$.

c Find in terms of $\mathbf{a}$ and $\mathbf{b}$: i $\overrightarrow{OP}$ ii $\overrightarrow{XP}$.

d What do your answers to c show about the
   points O, P and X?